Uncorking the Physics of Wine

Lutz Kasper · Patrik Vogt

Uncorking the Physics of Wine

A Wine Tasting in 50 Experiments

 Springer

Lutz Kasper
Schwäbisch Gmünd, Baden-Württemberg,
Germany

Patrik Vogt
Essingen, Rheinland-Pfalz, Germany

ISBN 978-3-662-68758-1   ISBN 978-3-662-68759-8   (eBook)
https://doi.org/10.1007/978-3-662-68759-8

This book is a translation of the original German edition "Physik mit Barrique" by Kasper, Lutz, published by Springer-Verlag GmbH, DE in 2022. The translation was done with the help of an artificial intelligence machine translation tool. A subsequent human revision was done primarily in terms of content, so that the book will read stylistically differently from a conventional translation. Springer Nature works continuously to further the development of tools for the production of books and on the related technologies to support the authors.

Translation from the German language edition: "Physik mit Barrique" by Lutz Kasper and Patrik Vogt, © 2022. Published by Springer Berlin Heidelberg. All Rights Reserved.

This Springer imprint is published by the registered company Springer-Verlag GmbH, DE, part of Springer Nature.
The registered company address is: Heidelberger Platz 3, 14197 Berlin, Germany

Paper in this product is recyclable.

Preface

Wine and physics—how do they go together? Where one is based solely on pleasure, the other may be associated with school trauma for some. Fine aromas and very relaxed situations on the one hand, chalk-dusty formulas and exam stress on the other. At least that's a widely held stereotype. But those who engage with it can also enjoy physics. With this book, we want to encourage our readers to rediscover the physics "behind things" or to see it from a new perspective and to enjoy it in a relaxed manner. We have chosen the topic complex "wine" because it offers a variety of connections to the world of physical phenomena in its richness of facets and not least because we appreciate the grape juice itself very much.

With this, we bring two ancient cultures together: the culture of wine with its several thousand-year-old history and the culture of physics, which can look back on over 2000 years, at least in its ancient philosophical roots and mechanical-practical origins. We are not the first to have seen such a connection. Already in the first century AD, Heron of Alexandria dedicated a significant part of his mechanical and pneumatic inventions to the decanting and refilling, portioning, and automated mixing of wine. We will discuss this later in the book. Some of the ideas compiled here are good old acquaintances from experimental physics lectures, while others are our own developments from our long-standing efforts to teach physics in real-life contexts at schools and universities.

The sequence of all the experiments presented here follows less the physical subject structure known from textbooks, such as mechanics, thermodynamics, etc. Instead, we try to orient ourselves on the course of a wine tasting or a successful evening with friends and good drinks. So we start with the opening of wine bottles in an astonishing methodical variety. Keep your corkscrew, but rest assured that in case of need, it can also be done with trees, kitchen torc, or bicycle pumps!

Once the bottle is opened, it often comes down to the correct aeration. Of course, one can make a meditative ceremony out of it. And some wines certainly deserve it. But what do you do when the guests arrive unexpectedly early? Decanting in under 60 s and emptying the bottle in 2 s could be a solution here. Aeration is usually followed by pouring, and should a drop spill, we have a scientifically proven excuse for you here.

We devote a larger part of the experiments to the complex gustatory perceptions that are obvious—or rather on the palate—in addition to the acoustic and optical phenomena. These are mainly offered to us by the "accessories", the glasses and bottles. We make them ring, sing and vibrate in resonance. We look in and through the glass, put on a red wine glasses and also show which tricks turn red wine into Blanc de Noirs or even water.

In a predominantly mechanical department, we become acrobatic and seemingly defy gravity. Floating corkscrews and glasses, or even the brimful spritzer glass in the somersault swing, will encounter you here. Or you can let Pythagoras himself teach you modesty.

At the end of the wine tasting, we may have run out of wine, but not the physical attention. We can still experiment with the remnants, the empty glasses and bottles, the candles, and the tablecloth.

You will see, a wine tasting is above all one thing: applied physics! Let yourself be inspired to experiment, to ask questions and give answers. And all this without any exam stress. All the experiments presented here have been successfully carried out by us several times. The vast majority of them can be carried out without special equipment. As a rule, just what is found on the table during a convivial evening is sufficient. For some measurements, we have used smartphone apps, which are available free of charge with few exceptions. However, as authors, we take no responsibility for any broken favorite glasses, unpredictable champagne corks, or red wine stains on clothing and furniture! In any case, we wish you an unclouded and significantly elevated wine enjoyment through physics and agree with Charles Darwin at this point: Only a fool doesn't experiment!

We thank Dr. Susanne Ihringer and Kevin Kärcher from PH Schwäbisch Gmünd for their helpful advice and experimental support on chemical issues.

Above all, we would like to finally thank our families and especially Kathrin and Diana for patiently enduring resonating bottles, singing glasses, cracking wooden slats, shattered glasses, and some strange arrangements on domestic tables, as well as for their great support in experimenting and for valuable feedback on texts or illustrations. The book would not have become what it is now without you, or not at all.

in March 2022 / May 2024

Lutz Kasper
Patrik Vogt

Contents

1 In the Beginning is the Cork: Physics of Bottle Opening 1

Uncorking Wine and Champagne Bottles 1

Experiment 1: "Pop"—How Fast is Sound? 1

Experiment 2: Uncorking by Striking the Bottle Bottom 6

Experiment 3: Uncorking for Ball Sports Players 8

Experiment 4: Uncorking with Heat—Slowly 10

Experiment 5: Uncorking with Heat—Quickly 13

Experiment 6: Pressure in Champagne Bottles 16

Experiment 7: Champagne in the Fog 19

Corkscrews—Sophisticated Machines 22

Experiment 8: The Corkscrew as a Load Crane 23

2 Well Ventilated? The Art of Decanting 25

Different Variants of Decanting 25

Experiment 9: Swirling—Old-Fashioned Gesture? 25

Experiment 10: The Surface Counts—What Lasts Long … 29

Experiment 11: Clever! Primitivo a la Venturi 31

Experiment 12: For Those in a Hurry—Flash
Aeration with the Smoothie Maker 33

For Those in a Hurry: Turbo Pouring with Pressure
and Tornado 35

Experiment 13: Pouring Time Halved Twice! 35

When the Wine Should not Breathe … 36

Experiment 14: Balloon and Vacuum Wine Pump 37

3   The Ear Drinks Too: Acoustics with Wine Glasses
    and Bottles   41
    Acoustics with Wine Glasses   41
        Experiment 15: Frequency of Wine Glasses   42
        Experiment 16: Cheers! Beat frequencies when
        clinking glasses   46
        Experiment 17: Shattering Wine Glasses?—Recording a
        Resonance Curve   49
    Sound Speed for Beginners and Advanced   53
        Experiment 18: Measurement at the "Palatinate Tube"   53
        Experiment 19: The Wine Glass as a Helmholtz Resonator   55
        Experiment 20: Wine Bottles as Helmholtz Resonators   57
    There's Music in It—Glasses and Bottles as Instruments   59
        Experiment 21: Wine Glass Harmonica   59
        Experiment 22: Wine Bottle Pan Flute   64

4   Enjoy with All Senses: Optical Phenomena with
    Wine Glasses   69
    Wine Glasses as Lenses   69
        Experiment 23: Burgundy Glass—Spherical Lens and
        Cobbler's Ball   69
        Experiment 24: Fun with Cylinder Glasses   73
    Looking Through Glass   75
        Experiment 25: Red Wine as a Color Filter   75
        Experiment 26: A Look into the Glass with
        the Infrared Camera   78

5   Of Good and Bad Drops: Fluid Dynamics of Wine   83
    Pouring? Yes Please, But Without Mishap!   83
        Experiment 27: Spilling as a Law of Nature   83
        Experiment 28: Carrying Wine with a Sieve?   86
        Experiment 29: Tears of Wine   89

6   The Well-Tempered Wine   91
    Frapper or Chambrer?   91
        Experiment 30: "Eco-Refrigerator" for a Wine Bottle   92
        Experiment 31: Wine Cocktail on the Rocks or:
        Is the Hugo Overflowing Now?   95
    A Digestif? Even when Frozen!   98
        Experiment 32: "Freezer Burn"   98

7   Magic Tricks and Wonders: Acrobatic Mechanics   103

Ancient Engineers—Water to Wine or Moderation?   103

  Experiment 33: Heron's Wine Automaton   104

  Experiment 34: With Pythagoras to Moderation   106

  Experiment 35: Thrifty by Diluting?   110

  Experiment 36: The Wine Glass as a Diving Bell   113

Balance Acts   116

  Experiment 37: Balancing Corkscrews   116

  Experiment 38: Bottle Holder at the Limit   119

  Experiment 39: The Floating Wine Glass   121

  Experiment 40: Spun, not Stirred …   124

  Experiment 41: The Falling Wine Glass   128

  Experiment 42: A Break Test for the Fearless!   131

8   Finished Drinking? Experimenting with Residual Alcohol   137

An Experiment with the Last Sip   137

  Experiment 43: The Paradoxical Wine Clock   137

Blow the Whistle! Acoustic Alcohol Test   140

  Experiment 44: Why Drunk People Whistle Higher   140

How does the Cork Get Out of the Empty Bottle?   144

  Experiment 45: Cork Release through Friction   144

  Experiment 46: Cork Release by Buoyancy   149

Does the Cork not Know Gravity?   151

  Experiment 47: Paradoxical Cork in a Glass of Water   151

Who Pays the Bill? An Experiment à la Otto von Guericke …   153

  Experiment 48: New "Magdeburg Wine Glasses"   154

Table, Cover Yourself … Up!   156

  Experiment 49: Tablecloth Away in 0.1 s   156

Bottle Empty? Lights Out!   159

  Experiment 50: Blowing Out Candles Through the Bottle   159

Bibliography and Apps Used   163

Index   167

List of Figures

Fig. 1.1	The bottleneck as a resonance tube	3
Fig. 1.2	Mouth correction on tubes	4
Fig. 1.3	Screenshot during cork pulling (App: Spectroscope)	4
Fig. 1.4	Wine bottle for "finger popping" with centimeter marks	5
Fig. 1.5	With each strike, the cork moves further out of the bottle neck	6
Fig. 1.6	Inserted ball needle (**a**) and connection of the bicycle pump (**b**)	8
Fig. 1.7	Uncorking a wine bottle with a bicycle pump and ball needle	9
Fig. 1.8	Uncorking a wine bottle with an air pump corkscrew	10
Fig. 1.9	Wine bottle in hot water bath (**a**) and thermal expansion of the liquid until the cork is pushed out (**b** to **d**)	11
Fig. 1.10	Thermodynamic model process when uncorking with the flambé burner	14
Fig. 1.11	Stroboscopic image of the movement (**a**), path-time diagram of the champagne cork (**b**)	17
Fig. 1.12	Prepared champagne cork (**a**), determination of the static friction force (**b**)	18
Fig. 1.13	Fog formation when opening a champagne bottle	20
Fig. 1.14	Pressure-volume diagram for the gas in the neck of a champagne bottle when opening	21
Fig. 1.15	Examples of different mechanisms of corkscrews	23
Fig. 1.16	Principle of the two-sided lever	23
Fig. 1.17	Principle of the one-sided lever	24
Fig. 1.18	Determination of the ratio "pulling force" to "lever force" on a wing corkscrew (left: the weight force on the two wings of 0.5 N each is not sufficient to lift the 1-kg piece. Middle: Here there is a force equilibrium between the two wings with 1.0 N each and the spiral with 10 N)	24

Fig. 2.1   Variant of the "Blue-Bottle-Reaction" in the swirled wine
            glass; before the swirling (**a**), after the swirling (**b**)   27
Fig. 2.2   "Blue-Bottle-Reaction" in the sealed flask   28
Fig. 2.3   Physical quantities determining the wave formation in
            the wine glass   29
Fig. 2.4   Different number of wave peaks at the same fill level due to
            variation of the angular velocity; only one wave peak (**a**), before
            the large wave peak on the right a smaller one can be clearly seen
            (**b**), at least four wave peaks (**c**)   29
Fig. 2.5   The same amount of wine in a decanter (**a**) and a water glass (**b**)
            immediately after opening the bottle   30
Fig. 2.6   Wine from decanter (**a**) and water glass (**b**), several days after
            opening the bottle   31
Fig. 2.7   Venturi pourer in action   32
Fig. 2.8   Aerating wine with a smoothie maker   34
Fig. 2.9   "High-speed techniques" for emptying a bottle   36
Fig. 2.10   Evidence of the negative pressure created by a vacuum wine pump
            in a wine bottle; commercial vacuum wine pump (**a**); sealed bal-
            loon with little filling in the bottle before pumping (**b**); inflated
            balloon after pumping (**c**); functional diagram of the pump with
            backflow valve (**d**)   37
Fig. 3.1   Sound analysis of a struck red wine glass   42
Fig. 3.2   Frequency spectrum of a struck red wine glass   43
Fig. 3.3   Oscillogram (**a**) and frequency spectrum (**b**) of the tone of
            a tuning fork   45
Fig. 3.4   Oscillogram (**a**) and frequency spectrum (**b**) of the sound of a
            piano (note A4)   45
Fig. 3.5   Quantitative analysis of the acoustic beat produced by two wine
            glasses   47
Fig. 3.6   Frequency spectra of the individual tones produced by the wine
            glasses   47
Fig. 3.7   Determination of the beat frequency   48
Fig. 3.8   Oscillograms of two beats; complete beat (**a**), incomplete beat (**b**),
            analyzed and displayed with the "Sound Analyzer" app   48
Fig. 3.9   Experimental determination of the natural frequency of the glass
            used   50
Fig. 3.10   Experimental setup for recording a resonance curve   50
Fig. 3.11   Time course of the amplitude with a stimulation at 614 Hz   51
Fig. 3.12   Graphical representation of the series of measurements (sound
            pressure level not calibrated)   52
Fig. 3.13   Determination of the speed of sound with the "Palatinate Tube"   54
Fig. 3.14   Screenshots for the measurement of the speed of sound; white
            noise without glass (**a**), resonance frequency in the "Palatinate
            Tube" (**b**) (App: Audio Kit)   54

Fig. 3.15 Determination of the speed of sound with bulbous glasses   56
Fig. 3.16 Historical (**a**) and unconventional Helmholtz resonators (**b–d**)   58
Fig. 3.17 Fundamental frequency of a blown wine bottle (App: Spaichinger
           sound analyzer); top: measured value, bottom: music note
           matching the measured frequency   59
Fig. 3.18 "Singing wine glasses" as a musical instrument   60
Fig. 3.19 Vibrations made visible; rubbed glass edge (**a**), struck glass (**b**) 61
Fig. 3.20 Comparison of the different sounds produced on the same
           glass with unchanged water filling; "rubbed" wine glass
           (**a**), struck wine glass (**b**)   62
Fig. 3.21 "Coffee mug model" for the fundamental vibration of
           a wine glass   62
Fig. 3.22 Glass harmonica with foot drive (CC-BY-SA 4.0, Historisches
           Museum Frankfurt (X25198), photo: Uwe Dettmar)   64
Fig. 3.23 Fill volumes, tones and fundamental frequencies of blown
           bottles using the example of a Bordeaux wine bottle   66
Fig. 4.1   The world in the glass is upside down   70
Fig. 4.2   A cobbler's ball as a historical craftsman's lamp   71
Fig. 4.3   Double refraction of divergent light through a spherical lens   71
Fig. 4.4   Refraction when parallel light enters a bulbous wine glass   72
Fig. 4.5   View through a cylindrical glass; glass directly in front of the
           inscription (**a**), glass somewhat removed from the inscription (**b**) 74
Fig. 4.6   Selected optical ray paths on a cylindrical lens   75
Fig. 4.7   Laser pointer with green and red light hitting red wine   77
Fig. 4.8   Various colored citrus fruits seen through a "red wine
           color filter"   78
Fig. 4.9   Even with most red grape varieties, the juice and the pulp
           of the berries are white, the colorants are mainly in the skin   79
Fig. 4.10 Sensitivity curve of commercial CCD chips   80
Fig. 4.11 Digital camera with IR filter (**a**), Infrared pass filter for screwing
           onto the camera objective (**b**)   80
Fig. 4.12 Red wine, taken with visible light (**a**), with infrared and visible
           light (**b**), only with infrared light (**c**)   81
Fig. 4.13 Sunset viewed through a wine glass   82
Fig. 5.1   Still image series of pouring from a wine bottle   84
Fig. 5.2   Contact angle ϑ between a liquid and a solid; hydrophilic
           (**a**), hydrophobic (**b**)   85
Fig. 5.3   The inverted glass holds tight! With a beer mat (**a**),
           with a sieve (**b**)   86
Fig. 5.4   Red wine forms typical "tears" in the glass after swirling   89
Fig. 5.5   Schematic explanation of the Marangoni effect in wine
           glasses (**a–c**)   90

Fig. 6.1	Comparative measurement of the effectiveness of a simple clay wine cooler	93
Fig. 6.2	Result of the measurement: The wine cooler does its job!	94
Fig. 6.3	Course of the melting process of an overhanging lump of ice in a glass filled to the brim	96
Fig. 6.4	Ice cube sinks in rum with 80% Vol. (**a**), ice cube floats in grain brandy with 28% Vol. (**b**)	97
Fig. 6.5	PET bottle filled with wine in freezer	99
Fig. 6.6	Decanting the ethanol and thawing part of the frozen wine	99
Fig. 6.7	Determination of the alcohol content before freezing (**a**) and after decanting (**b**) using a vinometer	100
Fig. 6.8	Determination of the alcohol content using a vinometer	101
Fig. 6.9	Capillary rise height and contact angle	101
Fig. 7.1	Heron's wine automaton, which turns water into wine (**a**); a wine jug that automatically refills a vessel (**b**)	104
Fig. 7.2	Photo (**a**) and diagram (**b**) of a Tantalus or Pythagorean cup	108
Fig. 7.3	Principle of the hydraulic siphon (**a**) and chain analogy (**b**)	109
Fig. 7.4	Wine mixing machine with preset mixing ratio based on Heron (cf. Schmidt, 1899)	111
Fig. 7.5	Principle of communicating tubes	112
Fig. 7.6	Wine glass as a diving bell	113
Fig. 7.7	Zinnowitz diving gondola on peninsula Usedom (Baltic Sea); a dive with up to 24 people lasts 30 to 40 min	114
Fig. 7.8	Relationship of pressure and remaining air volume of a diving bell at different water depths	115
Fig. 7.9	The cork with its "weights" continues to balance on the bottle neck even while pouring	116
Fig. 7.10	Centers of mass (CM) of the individual parts and the entire system	117
Fig. 7.11	The connected corkscrews continue to balance even after the stick has burned off (**a–c**)	117
Fig. 7.12	Bottle Holder "in action"; with a full (**a**) and an empty (**b**) bottle	119
Fig. 7.13	Centers of gravity in the bottle and holder system (**a**), forces, levers and torques on the wine bottle in the holder (**b**); force vectors are not to scale	119
Fig. 7.14	Right-hand rules for the vector product (**a**) and for the orientation of the direction of rotation for the torque (**b**)	121
Fig. 7.15	A wine glass appears to stand freely and without support on a glass plate	122
Fig. 7.16	Evidence of adhesive forces between water and glass	123
Fig. 7.17	The surface of a liquid accelerated in the direction of translation forms an angle to the horizontal, which depends on the amount of the acceleration	124

Fig. 7.18   Even with a complete rotation, the liquid surface remains
            approximately parallel to the base of the glass and
            nothing spills over   126
Fig. 7.19   Radial and tangential components of the acceleration measured
            with the accelerometer of a smartphone instead of the wine glass
            on the tray during several rollovers   126
Fig. 7.20   Free surfaces at rest (**a**) and during accelerated translational
            movement (**b**), based on Sigloch (2014)   127
Fig. 7.21   Setup of the described experiment (**a**) and modification with
            wine bottle, corkscrew and holding rod (**b**)   128
Fig. 7.22   Moment of the first overturn (**a**), stroboscope recording of the
            entire movement (**b**)   129
Fig. 7.23   At the beginning of a pirouette, the figure skater extends both
            arms and one leg (**a**); after pulling them towards the body, her
            angular velocity increases significantly (**b**)   131
Fig. 7.24   Consecutive images of a high-speed recording, taken at
            240 frames per second   132
Fig. 7.25   Hinge model of the breaking wooden strip   132
Fig. 7.26   The wooden strip initially lifts off the edge of the glass
            (top right), but then accelerates downwards overall due to the lack
            of breakage in the center   134
Fig. 8.1   Preparation of the experiment; one of the two shot glasses is
            completely filled with red wine (**a**), the other with water (**b**)   138
Fig. 8.2   The lighter red wine rises through the small opening and swaps
            places with the water   139
Fig. 8.3   Analogous experiment to the Gulf Stream; the water cooled by
            the ice cubes sinks, warm surface water flows from the left.
            A circulation occurs, which was made visible in the experiment
            by potassium permanganate   140
Fig. 8.4   Dog whistle used for the experiment   141
Fig. 8.5   Fundamental frequency of the whistle in relation to the blood
            alcohol concentration   142
Fig. 8.6   Deflection of one of the hung clothespins leads to the
            creation of a wave   144
Fig. 8.7   A bag almost effortlessly transports the cork out of
            the empty bottle   145
Fig. 8.8   Friction force between two bodies in general (**a**) and in the
            case of a bottle cork (**b**)   147
Fig. 8.9   Determination of the friction coefficients for cork-glass in a
            simple experiment   147
Fig. 8.10   Natural material "Cork"   148
Fig. 8.11   The cork is removed from the bottle belly by utilizing the
            buoyant force (**a–d**)   149

Fig. 8.12   Derivation of the buoyant force   150
Fig. 8.13   Floating cork slice; in the not full (**a**) and in the
            "overflowing" glass (**b**)   152
Fig. 8.14   Otto von Guericke: Experimenta nova (ut vocantur)
            Magdeburgica de vacuo spatio. (Engraving by Caspar
            Schott, 1672)   154
Fig. 8.15   Reconstruction of the "Magdeburg Hemispheres"
            experiment with wine glasses   155
Fig. 8.16   The tablecloth is cleared in 0.1 s   157
Fig. 8.17   Wine glass on pulled tablecloth; slowly pulled
            (**a**), pulled very quickly (**b**)   158
Fig. 8.18   Streamline image of an ideal flowing medium around a cylinder   160
Fig. 8.19   A burning candle is seemingly blown out through the bottle.
            The direction of the candle flame at the beginning of the
            process is astonishing   160

List of Tables

Table 1.1 Measurement results for "finger popping" and
calculated speeds of sound   5

Table 3.1 Frequencies of the fundamental tone and the
overtones of an A4 played on the piano   43

Table 3.2 Calculated and measured tuning of a bottle pan
flute for a temperature of 21 °C   66

Table 6.1 Optimal drinking temperatures of different types of
wine (see www.weinfreunde.de)   92

Table 8.1 Experimental determination of friction coefficients
($F_n = 4.61$ N in all cases)   159

1

In the Beginning is the Cork: Physics of Bottle Opening

Before we can begin—both with the wine tasting and with the experimenting—we have a sealed wine bottle in front of us. Usually, this poses no challenge and without paying much attention to the process, the cork is pulled out. But even this beginning of a wine tasting has its physical intricacies. Why is the cork stuck so tightly (we will not pay attention to bottles with screw caps in this chapter)? What causes the familiar sound when uncorking? What to do if no corkscrew is at hand? What pressure is actually in a champagne bottle? We will find answers to these questions in the first eight experiments and also take a look at the corkscrew as a force-saving machine.

Uncorking Wine and Champagne Bottles

Experiment 1: "Pop"—How Fast is Sound?

A still sealed wine bottle, a corkscrew and a smartphone are undoubtedly a promising combination. Notify good friends with one device and pull the cork out of the bottle with the other—what more could you want? But perhaps in addition to the throat, the "mind" is also thirsty and it should not be neglected here. So physical mindfulness is practiced right from the start, namely when opening the bottle. The characteristic pop sound when opening the bottle can give us information about the speed of sound (Kasper & Vogt, 2020). This is where the smartphone comes into play again. But first:

L. Kasper and P. Vogt, *Uncorking the Physics of Wine*, https://doi.org/10.1007/978-3-662-68759-8_1

Where does the pop sound actually come from and why does it sound like this? And finally: How can the speed of sound be measured with it?

The process of uncorking is accompanied by friction between the cork and the inner wall of the bottle neck as well as by airstreams between the cork edge and the bottle neck after their last contact. This creates a mixture of sounds from different frequencies. Now the situation in the bottle neck is comparable to that in some musical instruments, for example in a one-sided closed organ pipe. In acoustics, this is referred to as so-called *stopped pipes*. From a physical point of view, such an organ pipe is nothing more than a one-sided closed resonance tube. The air in such tubes vibrates preferentially at very specific frequencies and thus produces a specific sound. This sound depends mainly on the length of the organ pipe and less on its diameter. Now the neck of a wine bottle with the small amount of residual gas in it (strictly speaking, it is air with alcohol vapor components) is not an organ pipe, but the physics works according to the same principles. From the multitude of sound frequencies that arise when uncorking, the bottle neck "selects" its preferred ones, resulting in a specific frequency spectrum of the fundamental tone and the overtones. The brief moment of uncorking and the strong damping of the vibration make the resulting sound fade very quickly. This is how the very short and distinctive pop sound is created when uncorking.

If you have a suitable app on your smartphone or on your computer and let the bottle neck resonator preferred frequency (resonance frequency) simply measure (see used apps in the bibliography). A measurement example recorded with the smartphone is shown in the screenshot in Fig. 1.3. The very clearly recognizable peak value is exactly our sought resonance frequency.

Bottle Neck and Organ Pipes—An Acoustic Comparison

One-sided closed Organ pipes ("stopped pipes") have at their closed end a node of motion (and thus a pressure antinode) and at the open end an antinode of motion (and thus a pressure node). For the fundamental vibration in the case of resonance, this results in a quarter wavelength fitting into the reso nance tube (dashed curve in Fig. 1.1). From the general relationship of frequency f, wavelength λ and propagation speed $c = \lambda \cdot f$ the frequency f_0 of the fundamental vibration is obtained:

$$f_0 = \frac{1}{4L} c_{gas}$$

(L: Length of the gas column in the bottleneck; c_{gas}: Speed of sound in the gas).

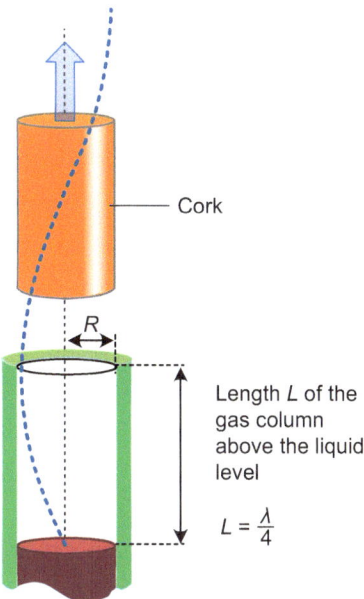

Fig. 1.1 The bottleneck as a resonance tube

For the residual gas in the bottleneck, the simplifying assumption should be made that it is air, disregarding the alcohol vapor components. Thus, c_{gas} can be replaced by c_{air}. However, for a more accurate estimate of the speed of sound another effect should be considered. The plane of the antinode of motion does not exactly coincide with the plane of the opening at the bottleneck. This can be explained by the fact that the air particles in the plane of the bottle opening start to vibrate and the pressure node (the place of minimal pressure fluctuations) shifts slightly outward (Fig. 1.2). For this reason, a length correction, the so-calledend correction ΔL, is required. This corresponds exactly to the distance between the planes of the bottle opening and the shifted pressure node.

The end correction depends on the radius of the opening. We know its value from measurements (cf. Levine & Schwinger, 1948):

$$\Delta L = 0.61R$$

This gives us the speed of sound when uncorking:

$$c_{air} = 4f_0(L + \Delta L)$$

In the measurement example, the resonance frequency is 1254 Hz. This value is needed to determine the speed of sound according to the equation formulated in the info box. Now the length of the gas column from the wine level to the top edge of the bottle neck must be measured. For better

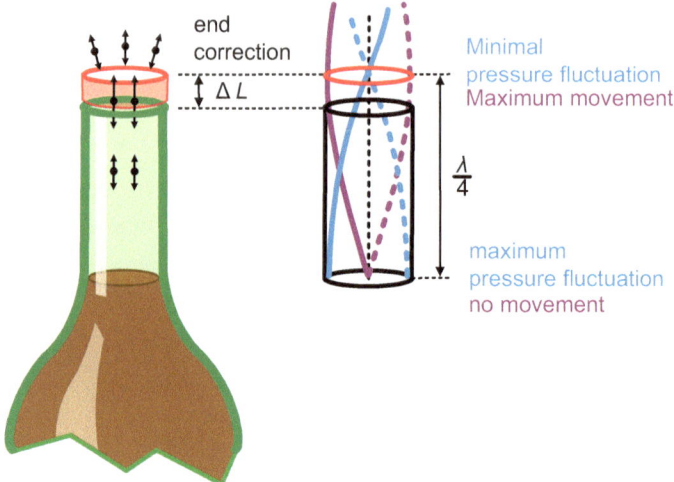

Fig. 1.2 Mouth correction on tubes

Fig. 1.3 Screenshot during cork pulling (App: Spectroscope)

accuracy of the speed of sound determination, the diameter or radius of the bottle neck is finally measured after uncorking.

The example measurement from Fig. 1.3 was carried out on a wine bottle that has a 6 cm long gas column. The inside measured radius of the bottle neck opening is—as with most standard wine bottles—one centimeter. Thus, the speed of sound can now be estimated. If all values are inserted into the equation for the speed of sound, a value of 332 m/s is obtained. The ambient temperature of 23 °C determined for this experiment theoretically

Fig. 1.4 Wine bottle for "finger popping" with centimeter marks

Table 1.1 Measurement results for "finger popping" and calculated speeds of sound

Length of the gas column L in m	Measured resonance frequency f_0 in Hz	Calculated speed of sound c_{air} in m/s
0.030	2355	340
0.040	1740	321
0.050	1488	334
0.060	1255	324
0.070	1058	322
0.080	925	328

allows a speed of sound of 345 m/s and is therefore actually close to the measured value.

With the emptied bottle, further systematic experiments can then be carried out—possibly only the next day. Although the cork is irrevocably out, instead of uncorking, the sound can be reproduced by "popping out" from the bottle neck with a moistened finger (Fig. 1.4 and Table 1.1).

Now we can repeat the measurement on the water-filled bottle at various fill levels. One advantage of this measurement is that there is actually only air in the bottle neck. The table shows the results of such a systematic series of measurements. The relative error is also in the order of 5%. Not bad for a start! During the wine tasting, we will return to the sound speed measurement and it is already promised: We can do it even more accurately! But first, we stay with the topic "Corkscrew" and provide first aid in the next experiment for the emergency that no corkscrew is at hand …

Experiment 2: Uncorking by Striking the Bottle Bottom

In this and the following experiments, we want to investigate how a wine bottle can be opened without the use of a corkscrew. Imagine, for example, that you have prepared a nice picnic for a hike and then actually have neither a corkscrew nor a pocket knife at hand. What at first glance appears to be a very unfortunate situation is actually not so problematic. Without knowing it, you are actually carrying a corkscrew with you in most situations. This is by no means as physically well thought out as the variants described in Experiment 8, but still practical. We are talking about your shoes! How is that supposed to work, you ask? Quite simple! You put the bottle of wine in the shaft of your shoe and strike the shoe and bottle several times hard against a solid surface (Fig. 1.5). This can be the ground, a wall

Fig. 1.5 With each strike, the cork moves further out of the bottle neck

or in the forest also a tree trunk, and with each strike, the cork moves a few millimeters out of the bottle neck. At some point, the cork protrudes so far that it can be pulled out by hand, and the picnic can begin.

What is the explanation for the movement of the cork? While the bottle is struck by hand against the surface, the wine of course also moves in the direction of the impact. Although the bottle is cushioned by the sole of the shoe during the impact, it comes to rest quickly. However, the wine wants to continue moving and is reflected like a ball thrown against a wall from the bottle bottom. Thus, the liquid sloshes in the other direction and strikes against the cork. This force increases the momentum of the cork and sets it in motion. The movement comes to a halt quickly due to the acting sliding friction force between glass and bottle neck, which is why the cork moves only a few millimeters per strike.

At this point, we would like to point out that the experiment should be carried out with a lot of force, but still with all caution.

Momentum and Impulse

The momentum describes the state of motion of a body taking into account its mass m and velocity $\vec{v}$. It is proportional to both quantities and the following applies:

$$\vec{p} = m \cdot \vec{v}$$

In everyday life, we sometimes use terms like "force" or "verve" for the momentum of a body. To be distinguished from momentum is the impulse $\vec{I}$, which describes the effect of a force $\vec{F}$ on a body during the time t and leads to a change in momentum:

$$\vec{I} = \int_{t_1}^{t_2} \vec{F}\, dt = \Delta \vec{p}$$

If the acting force and the mass of the body are constant, the relationship simplifies to:

$$\vec{F} \cdot \Delta t = m \cdot \Delta \vec{v}$$

The application of a force over a certain time thus leads to a change in velocity of the considered body. In the experiment described here, the resulting impulse leads to a short-term increase in cork speed.

Experiment 3: Uncorking for Ball Sports Players

The following variant of bottle opening can only be described as absolutely spectacular! It is particularly suitable for readers who play a ball sport, such as handball, volleyball or basketball. A bicycle pump with a ball needle attachment is used, and the procedure is extremely simple.

Since the ball needle is usually a bit shorter than a cork, we first pierce it with a thin pointed object, e.g., a fine knitting needle. Now the ball needle is inserted into the pre-made channel and a bicycle pump is connected (Fig. 1.6).

Now we start pumping and even though the ball needle does not completely pierce the cork, air can flow into the bottle through the prepared channel. This leads to an increase in pressure and thus to an increase in the force acting on the cork from below (see info box). If we pump hard enough—in fact, only one or two strokes are required—the frictional force acting between the cork and the bottle neck is overcome, and the cork begins to slide (see info box of Experiment 45, Fig. 1.7).

From what pressure the uncorking starts, we can read off the pump's manometer and thus estimate the frictional force acting between the cork and the bottle neck. In our measurement example, the cork started moving at a displayed pressure of about 7 bar, which corresponds to the following force for an inner radius of the bottle neck R of 1 cm:

$$F = p \cdot A = p \cdot \pi R^2 = 700,000 \, \text{Pa} \cdot \pi \cdot (0.01 \, \text{m})^2 \approx 220 \, \text{N}$$

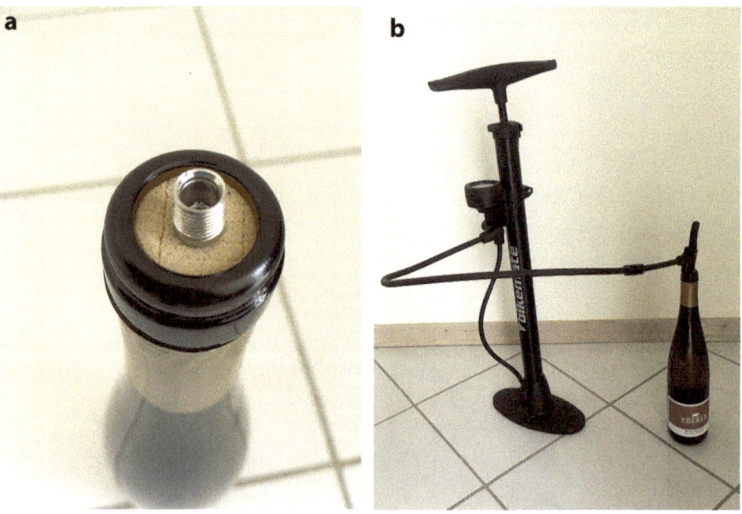

Fig. 1.6 Inserted ball needle (**a**) and connection of the bicycle pump (**b**)

Fig. 1.7   Uncorking a wine bottle with a bicycle pump and ball needle

(*A:* Cross-section of the bottle neck). For the bottle used in the experiment, the frictional force between glass and cork is therefore about 220 N. Due to the very simple manometer, this is initially only a rough estimate, which can however be well confirmed by the measurement series of Experiment 5. It should also be noted that the pump manometer displays the excess pressure in the bottle relative to the atmosphere. The absolute pressure is therefore about 1 bar higher, which however does not matter for the estimation of the frictional force.

The method described here is actually also used in many commercially sold corkscrews. There have long been bottle openers that work with a cannula and gas cartridge. However, recently, discount stores have increasingly been offering "air pump wine openers", i.e., our bicycle pump in small (Fig. 1.8).

Definition of Pressure

The physical quantity pressure *p* corresponds to the quotient of the force *F* acting perpendicularly and the area *A:*

$$p = \frac{F}{A}$$

Pressure thus indicates the force acting on an area of one square meter. Its SI unit is the Pascal (1 Pa), named after the French mathematician, physicist and philosopher *Blaise Pascal* (1623–1662). More commonly used in everyday life is the unit bar (1 bar), where 1 bar = 100,000 Pa.

Fig. 1.8   Uncorking a wine bottle with an air pump corkscrew

Experiment 4: Uncorking with Heat—Slowly

Is repeatedly hitting the bottom of the bottle too violent for you? And you're still missing the corkscrew? In this case, we suggest a calm, almost meditative variant of "cork pushing" instead (it's certainly not corkscrewing). However, for this relaxed way of opening the bottle, a little time must be planned and the disadvantage of a significantly too warm wine must be accepted.

Apart from a stove and a pot with some water, no further utensils are needed. The pot with water is placed on the hot stove and the wine bottle is placed in it so that at least a third of the bottle is covered with water (Fig. 1.9a). The red mark on the right in the picture indicates the wine level in the bottle neck at normal room temperature. The stove is now heated to the maximum until the water begins to boil. Then it can be turned down so that it continues to simmer. Now it takes some time. Anyone interested in the physics of this type of uncorking should observe the processes closely—an important virtue when experimenting!

Fig. 1.9 Wine bottle in hot water bath (**a**) and thermal expansion of the liquid until the cork is pushed out (**b** to **d**)

After a few minutes, a change in the wine level in the bottle neck can be seen (Fig. 1.9b and c). Quite obviously, the wine expands. Since no change can be seen at the position of the cork, we can conclude that the small volume of gas between the level and the bottom of the cork is being reduced. Unlike liquids, which can hardly be compressed, gases can be easily compressed. With stoic calm, the wine follows the laws of thermodynamics and its own expansion coefficient (see info box) and continues to expand.

At a certain point of heating, the wine level—separated only by a very small strongly compressed gas layer—reaches the cork and the static friction force of the cork, which has been firmly seated until then, is finally overcome. The cork is gently but firmly pushed out by the further expanding wine (Fig. 1.9d). From now on, the effect of expansion on the cork can be observed even better. Since only the lower third of the bottle is in the hot or then also boiling water in our experiment, it takes some time to heat the entire contents of the bottle. In an experiment we conducted, it took 14 min from the start of boiling in the pot until the cork was completely pushed out of the bottle!

In this experiment, the wine bottle can ultimately be compared to a large liquid thermometer. Each of these thermometers has a small ball- or cylinder-shaped reservoir for the thermometer liquid (often alcohol) at the lower end of the scale, to which the capillary tube, a narrow capillary, is attached. Basically, our wine bottle with its large reservoir and the "bottle neck capillary" is similarly constructed. Also, there is ethanol in our bottle thermometer, even if it is only about 13% Vol.

By the way, anyone who is worried about the boiling water and the resulting high temperatures possibly endangering the alcohol content of the wine can be reassured here. Although the boiling point of ethanol is already reached at 78 °C and it would evaporate quickly under normal conditions.

The temperature in the lower area of the bottle is probably exceeded. However, only a minor evaporation of the ethanol actually occurs, as the bottle content is under high pressure due to the liquid expansion. Already at an increase of pressure to one and a half times the atmospheric pressure, the boiling point of ethanol is already 90 °C. However, a much higher pressure is required to move the cork (see Experiment 3 and 5) and the boiling point for the alcohol can therefore no longer be reached.

Spatial Expansion Coefficient

Solid, liquid and gaseous substances generally change their dimensions with temperature changes. This behavior characterizes the substances and is referred to as spatial (cubic) expansion coefficient γ. The temperature dependence of the volume of a substance is given by:

$$V(T) = V_0(1 + \gamma \Delta T)$$

(ΔT: temperature difference; V_0: initial volume before temperature change)

The expansion coefficient itself can also be temperature-dependent. This effect is particularly significant in water. The negative value of the expansion coefficient for water between 0 and 4 °C is referred to as *anomaly of water*. In this range, the volume decreases with increasing temperature.

Examples of expansion coefficients:

Ethanol (20 °C)	0.0011 K^{-1}
Water (0 °C)	-0.000068 K^{-1}
Water (20 °C)	0.000206 K^{-1}
Water (100 °C)	0.000782 K^{-1}

By the way, the final temperature of the wine when opening with the "slow" heat method can be relatively easily estimated. For this, the spatial expansion coefficient γ for water and ethanol the expansion coefficient for wine is calculated. For the water, due to the large temperature dependence, a mean expansion coefficient $\overline{\gamma}$ for the expected temperature range of about 20 to 70 °C must be assumed. We have determined this experimentally: $\overline{\gamma}_{\text{water}}(20\ldots70\,^{\circ}\text{C}) = 0.00043\,\text{K}^{-1}$.

For ethanol, there is also a temperature dependence, but it is significantly smaller than in water and we use the value given for 20 °C as a simplification (see info box).

This results in an ethanol volume fraction in the wine of 13%, the average expansion coefficient of the wine for the given temperature range:

$$\gamma_{\text{wine}}\left(20\ldots70\,^{\circ}\text{C}\right) = 0.13 \cdot \gamma_{\text{ethanol}} + 0.87 \cdot \overline{\gamma}_{\text{water}} \approx 0.00052\,\text{K}^{-1}$$

Next, we need to know by what volume the wine has expanded. This is simple, as the wine level at room temperature was about 6 cm below the opening of the cylindrical bottle neck. The bottle neck has an inner radius of 1 cm. Thus, the volume change is ΔV:

$$\Delta V = L\pi R^2 = 0.019\,1$$

Now we have all the necessary values and rearrange the equation of the temperature-dependent volume (see info box) for the desired temperature:

$$\Delta T = \left(\frac{V(T)}{V_0} - 1\right)\gamma_{\text{wine}}^{-1} = \frac{\Delta V}{V_0 \cdot \gamma_{\text{wine}}} = \frac{0.019\,1}{0.75\,1 \cdot 0.00052\,\text{K}^{-1}} \approx 49\,\text{K}$$

At a room temperature of about 22 °C, we can therefore assume an average wine temperature of around 71 °C for the moment when the cork is completely displaced from the bottle neck. This estimate contains some simplifications. In addition to the assumption of the average expansion coefficient, this also includes the lack of consideration of the pressure conditions in the bottle before the cork is pushed out. Nevertheless, the thus estimated temperature of the wine can be well confirmed by our measurement.

Even though we have seen that we do not have to worry about the evaporation of the alcohol at this temperature, it is still the case that the entire bottle content is heated in the water bath. If you do not intend to drink a mulled wine, you should urgently find another way to open the bottle for the sake of the wine. We will pursue such an alternative, which also works "thermodynamically", in the next experiment.

Experiment 5: Uncorking with Heat—Quickly

The thermodynamic opening of the wine bottle can of course also be done much faster and is then much gentler on the precious content. Admittedly: Anyone who owns a flambé burner in the kitchen usually also has a proper corkscrew. But maybe you are just about to flambé meat or vegetables and want to offer your guests a wine as well. And if you already have the burner in your hand …

In addition, we will pursue this method because it is also very appealing for physical reasons.

An unopened wine bottle always has a reservoir of gas between its tightly fitting cork and the wine level, which consists mainly of air and some water and alcohol vapor. If the bottle neck is strongly heated from the outside at this point—for example with a flambé burner (Fig. 1.10), then after about one minute you can actually experience how the cork pushes out of the neck. And all this without any effort! Of course, we pay for this with the energy from the gas of the burner.

But why does this work at all? An obvious assumption would be the expansion of the trapped air due to intense heating. Let's estimate how much the trapped air would need to heat up to push out the cork: An average wine bottle, after opening, has a wine level about 6 cm below the opening. With a 4 cm long cork, there is therefore a reservoir of 2 cm height. The inner radius at the neck of standard wine bottles is almost always about 1 cm. To push the cork completely out of the bottle, the gas volume must therefore approximately triple. What temperature is required for this? The basic laws of thermodynamics show us that we must consider a third variable, namely the pressure. All three variables together characterize the thermodynamic state of a gas and are described using the so-called ideal gas equation.

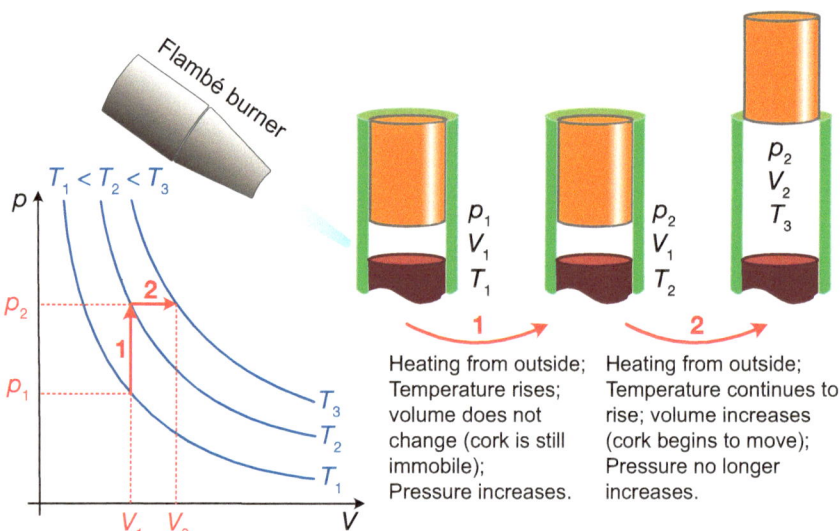

Fig. 1.10 Thermodynamic model process when uncorking with the flambé burner

---

Equation for an ideal gas

Changes in state of gases can be described by the following relationship of the three variables pressure p, volume V and temperature T:

$$\frac{p_1 \cdot V_1}{T_1} = \frac{p_2 \cdot V_2}{T_2}$$

Here, the indices 1 and 2 represent an initial and final state. Strictly speaking, the *ideal gas equation* only applies to so-called ideal gases, for which special assumptions are made. However, for estimates with gases like air, this equation can also be used as an approximation.

Upon closer examination, the "cork pulling" with the burner becomes a "cork pushing" by the trapped gas. This is a fair deal: From the outside, the gas in the bottle neck is supplied with energy by the burner, and in return, it does the hard work for us. The cork only starts to move when the pressure is high enough to exert a corresponding force on the firmly seated cork. What magnitude of force are we actually talking about here?

In a series of experiments, it was determined that the required force when pulling a cork is almost always over 250 N. This corresponds to the weight of a mass of over 25 kg! No wonder that ingenious people have come up with sophisticated "cork pulling machines" (which we will discuss in more detail in Experiment 8).

So, a force F of at least 250 N must act on the lower surface of the cork for it to start moving. The area of the bottom of the cork A is obtained from the square of the radius multiplied by the number π. With this, we can use the equation for pressure known from school physics (see info box for Experiment 3):

$$p = \frac{F}{A} = \frac{250\,\text{N}}{\pi \cdot (1\,\text{cm})^2} \approx 8 \cdot 10^5\,\text{Pa} = 8\,\text{bar}$$

If we assume that the pressure in the bottle neck is approximately equal to the atmospheric pressure of 1 bar, then eight times the pressure is required to set the cork in motion. Until this happens, of course, the gas volume in the bottle neck remains unchanged. This means that tripling the volume is not needed because the pressure remains approximately constant from the moment the cork starts moving (Fig. 1.10). With this, we have all the information to find out the required temperature of the trapped gas to push the cork. We use the gas equation and can cancel out the constant volume and rearrange for the required temperature:

$$T_2 = \frac{p_2}{p_1} \cdot T_1 = 8 \cdot 295 \,\text{K} = 2360 \,\text{K} \approx 2087\,°\text{C}\; (!)$$

The result of a temperature of over 2000 °C is surprising! According to the manufacturer, the blowtorch only reaches a maximum temperature of "only" 1300 °C.

Thus, we need to supplement our initial assumption. The expansion of the gas volume in the bottleneck alone cannot lead to the cork being pushed out. A second effect must come to the rescue here. And that can only be the conversion of the wine into steam at the surface. Boiling under increased pressure is something that is known from cooking with a pressure cooker. The temperature necessary for boiling becomes higher the higher the pressure is. The same applies when opening a bottle with a burner. That the pressure here is about 8 bar, which is significantly higher than in a pressure cooker, where a maximum of 1.8 times the normal air pressure is reached. There, water boils at 117 °C. In the bottleneck at 8 bar, it is already 170 °C! This temperature must at least be reached at the level surface in the bottleneck if the cork is to be loosened. For the alcohol content in the wine (approx. 10 to 14% Vol.), this temperature is lower and is approx. 130 °C.

With our next experiment, we will stay a little longer with the topic "Pressure", which is hopefully present in sparkling wine even without a temperature increase.

Experiment 6: Pressure in Champagne Bottles

After we have already opened five different variants of wine bottles, we would like to turn our attention to champagne bottles in this experiment. Everyone knows it, the starting shot of most family celebrations, which promises a tingling experience on the tongue and whose acoustics we have already addressed in Experiment 1. Of course, we are talking about the "pop sound", which goes hand in hand with the shooting out of the champagne cork and apparently requires much less effort than with wine bottles. But how does it happen that champagne bottles almost uncork themselves after removing the agraffe—this is what the wire mesh that fixes the champagne cork until opening is called? The cause of this is the excess pressure present in the champagne bottle, which can be traced back to the carbon dioxide produced during fermentation. And of course, there is also a regulation in Europe that prescribes to the winemaker how high this excess pressure must be at least so that he can call his product "quality sparkling wine"—that would be the correct term for champagne. In the corresponding European regulation (Official Journal of the European., 2008), it says in Annex IV, paragraph 5c: "Quality sparkling wine is the product that has an overpressure of at least 3.5 bar due to dissolved carbon dioxide in closed containers at 20 °C." In the following, we would like to introduce a simple method by which you can check whether your favorite champagne actually complies with this regulation (Vogt & Kasper, 2015). The autonomous shooting out of the cork is recorded with the high-speed mode of a digital camera and the video is analyzed with suitable software.

As a rule, the pressure prevailing in the bottle is not sufficient to overcome the static friction force between the cork and the bottleneck. Therefore, the cork is slightly loosened by hand after the start of the video recording and the end of the bottleneck is marked on the plastic cork with a waterproof pen.

The result of a video analysis carried out is shown in Fig. 1.11. On the left is a stroboscopic image of the movement, on the right a path-time diagram, the underlying data of which were obtained with the "measure Dynamics" software. The cork needed just 4 ms for the first 6 cm, resulting in the average acceleration a of 7500 m/s². This is an extremely large value and corresponds to about 750 times the acceleration due to gravity! After this acceleration phase, the cork continues to move at about 26 m/s, which is a speed of around 94 km/h! With the experimentally determined acceleration,

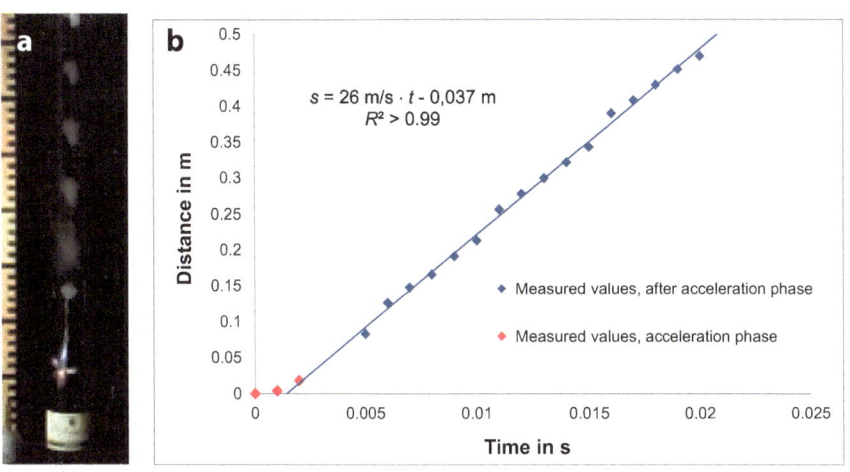

Fig. 1.11 Stroboscopic image of the movement (**a**), path-time diagram of the champagne cork (**b**)

the mass m of the champagne cork (0.007 kg) and the inner radius of the bottleneck ($R = 9.4$ mm), we can already make an estimate of the pressure and it applies (see info box of Experiment 3):

$$p_1 = \frac{F}{A} = \frac{m \cdot a}{\pi \cdot R^2} \approx 2 \, \text{bar}$$

According to this estimate, the excess pressure in the bottle is 2 bar, and the EU regulation would not be met. However, our calculation is very simplified and we have only estimated the excess pressure necessary for the acceleration of the cork. In addition, the frictional force between the bottle neck and the cork must also be overcome, so the actual excess pressure is higher.

To determine the frictional force acting between the bottle neck and the cork, the plastic cork was pierced and a hook with a lock nut was attached according to Fig. 1.12a. (By the way, this is also the reason why we used an industrial sparkling wine with a plastic closure, although we would actually prefer a nice winemaker's sparkling wine!) To determine the static friction force present when the bottle is uncorked by itself, the cork is inserted up to the mark and uncorked by pulling it out vertically with a suitable force meter (e.g., 250 N) (Fig. 1.12b). This resulted in a static friction force of $F_R = 75$ N. To overcome this force, the excess pressure p_2 is needed:

$$p_2 = \frac{F_R}{A} = \frac{75 \, \text{N}}{\pi \cdot \left(9.4 \cdot 10^{-3} \, \text{m}\right)^2} \approx 2.7 \, \text{bar}$$

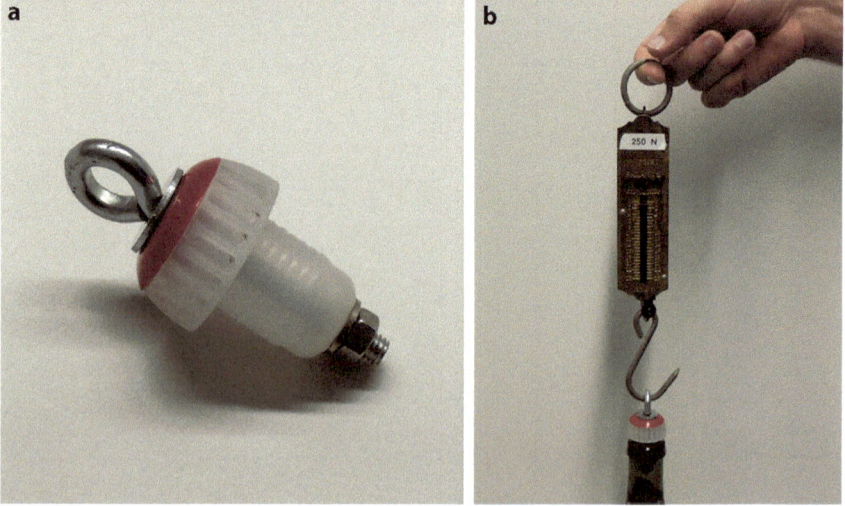

Fig. 1.12 Prepared champagne cork (**a**), determination of the static friction force (**b**)

To determine the excess pressure present in the champagne bottle, we finally form the sum of p_1 and p_2 and obtain approximately 4.7 bar, which is very well confirmed by the experimental work of Liger-Belair et al. (2017). This means that the 3.5 bar required for sparkling wine is even significantly exceeded!

How can we visualize a pressure of 4.7 bar? This pressure is present, for example, at a water depth of about 40 m or when about 50 small cars are stacked on an area of one square meter.

Acceleration and Force

The acceleration a describes the change in motion. For simplification, let's consider a one-dimensional motion, which corresponds to the change in velocity Δv per time interval Δt.

$$a = \frac{\Delta v}{\Delta t}$$

The SI unit of acceleration is therefore 1 m/s². If the acceleration of a vehicle is, for example, 3 m/s², then the speed of the vehicle increases by 3 m/s per second.

For a body of mass m to change its state of motion, a force F is required. The acceleration is proportional to the acting force and the following applies:

$$F = m \cdot a$$

This is the basic law of mechanics, which goes back to *Isaac Newton* (1643–1727). The SI unit of force is named after him and is called Newton (1 N).

Experiment 7: Champagne in the Fog

Did you pay close attention to what happened when you opened the champagne bottle in the last experiment? If not, you may have no choice but to open a second bottle. However, you may have already noticed the phenomenon: The moment the cork leaves the bottle, a clearly visible whitish-gray fog forms above the bottle neck (Fig. 1.13). Almost as quickly as it formed, the fog also dissipates.

How does this phenomenon occur and what is the fog made of? The only possible components are the gases that are between the liquid level and the cork in the bottle neck. These are air, additional carbon dioxide gas from the sparkling wine, alcohol vapor, and water vapor.

In our experiment, we assume the normal case and take the champagne from the refrigerator, at a temperature of about 6 °C, which also corresponds to the recommended serving temperatures (Table 6.1). In the still sealed

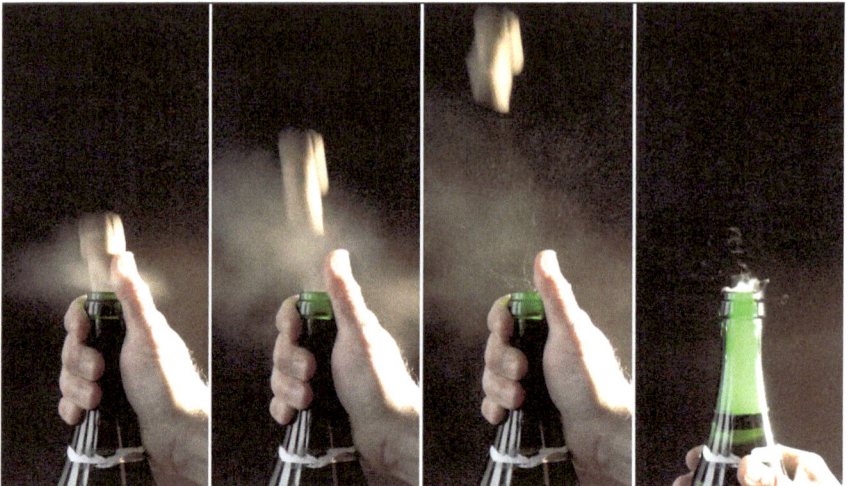

Fig. 1.13 Fog formation when opening a champagne bottle

bottle neck, there is then a pressure of about 4.7 bar, as we have shown in the previous experiment. Outside the bottle, we can assume the normal atmospheric pressure of about 1 bar. This pressure difference is what pushes the cork against the acting friction force out of the bottle neck. The exciting moment when the flying cork finally releases the opening is associated with a dramatic and very rapid drop in pressure in the bottle neck. The time span is too short for there to be a heat transfer with the expanding gas and its surroundings. In physics, such rapid state changes of a gas without thermal energy transfer are referred to as *adiabatic state changes* (see info box). In our case, the gas volume in the bottle neck suddenly expands into the outside space, while the pressure drops rapidly. In Fig. 1.14, it can be seen that the temperature decreases during this process. Measurements by Liger-Belair et al. (2017) have shown that it drops to almost −80 °C when opening a champagne bottle. At this temperature, very small ice crystals form from the water vapor, which we can see as fog. As the starting temperature of the bottle increases (e.g., at room temperature), the proportion of CO_2 gas and thus the pressure in the bottle neck also increases. If such a bottle is opened, the higher pressure difference to the atmospheric pressure during the adiabatic expansion results in an even stronger cooling, so that increasingly the CO_2 gas also freezes into ice crystals. This is known as dry ice, which quickly sublimates back into the gaseous state (sublimates), making the fog disappear quickly.

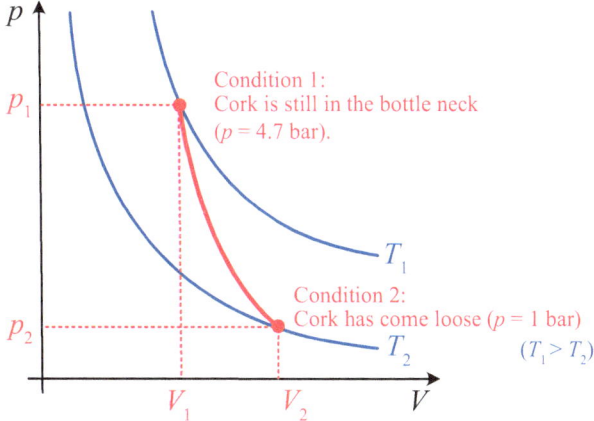

Fig. 1.14 Pressure-volume diagram for the gas in the neck of a champagne bottle when opening

Adiabatic State Changes

Gases in physics are described by the three essential quantities pressure p, temperature T and volume V. We consider here a quantity of gas as a closed system, in which no material transport across the system boundaries, but heat exchange and mechanical work can occur. The state of a gas can then be described with the general gas equation:

$$\frac{p \cdot V}{T} = \text{constant} \Rightarrow \frac{p_1 \cdot V_1}{T_1} = \frac{p_2 \cdot V_2}{T_2}$$

Often, states and state changes of a gas are given in $p(V)$-diagrams (Fig. 1.14). The temperature curve in this diagram results from the relationship: $p \cdot V = \text{constant} \cdot T$.

In Fig. 1.14, these are the blue curves. There, the following applies: $T_1 > T_2$. The state change between two points "State 1" and "State 2" can basically be carried out in different ways. For example, it would be conceivable in the diagram in Fig. 1.14 to conduct the process in such a way that starting from state 1 at constant temperature (i.e., along the T_1 curve) the gas expands to the volume V_2, where the pressure would decrease to a value between p_1 and p_2. Such a state change at constant temperature is called *isothermal*. Subsequently, the pressure could be reduced to the target value p_2 while keeping the volume constant at the value V_2. This would also lower the temperature from T_1 to T_2. Such a state change at constant volume is called *isochoric*. Both cases, isothermal and isochoric state change, are associated with energy exchange processes of the gas with its surroundings. However, the gas has no time for this exchange when opening a champagne bottle. The process takes place in the range of milliseconds. Such processes without heat exchange are called *adiabatic state changes* (red curve in the diagram in Fig. 1.14). Opening a champagne bottle leads to an adiabatic expansion, where the temperature decreases. The expansion work of the gas is "paid" from the internal energy of the gas, whose temperature then decreases very quickly.

Adiabatic processes often occur in our natural and technical environment. An example from acoustics are the very fast pressure fluctuations in the propagation of sound in gases.

Corkscrews—Sophisticated Machines

Do you know this: You want to open a wine bottle, turn the spiral of the tool into the cork, pull as hard as you can, and still hardly get the thing out of the bottle? Probably then you have used a very simple corkscrew. With it, a force may be required that corresponds to lifting a mass of up to 30 kg!

Fortunately, ingenious engineers have taken on these tools and have a multitude of physical laws working for us. Proper little machines can be corkscrews. Bells, spindles, screws, gears or even lever arms can be found in and on them. In physics, such things are called *force-transforming devices*. They mainly reduce our force effort through single and double-sided levers as well as inclined planes. These are the threads of the spindles on better corkscrews (but not the spirals that we screw into the cork). We buy the advantage of less force effort with an extra effort of covered lifting or turning paths. In physics, this "trade" is also called the *Golden Rule of Mechanics* (see info box). Fig. 1.15 shows some examples of technical implementation. Besides, the imagination of the developers knows almost no bounds. In the scientifically demanding corkscrew even more sophisticated principles come into play. One strategy is to reduce the friction between the cork and the glass wall, another idea is to increase the gas pressure under the cork against the external air pressure by means of a pierced cannula until it finally pushes out the cork (see Experiment 3).

Corkscrews as force-transforming devices

In corkscrews, the laws of levers are often used to reduce the necessary force. Basically, two types of levers are used, the *two-sided* lever (Fig. 1.16) and the *one-sided* lever (Fig. 1.17). For both types of levers, the so-called *Golden Rule of Mechanics* applies: What you save in force, you have to add in distance. This refers to the "stroke". With respect to the lever arm lengths, the equality of the mathematical product of the corresponding lever arm lengths (L_1; L_2) and forces (F_1; F_2) applies.

$$F_1 \cdot L_1 = F_2 \cdot L_2$$

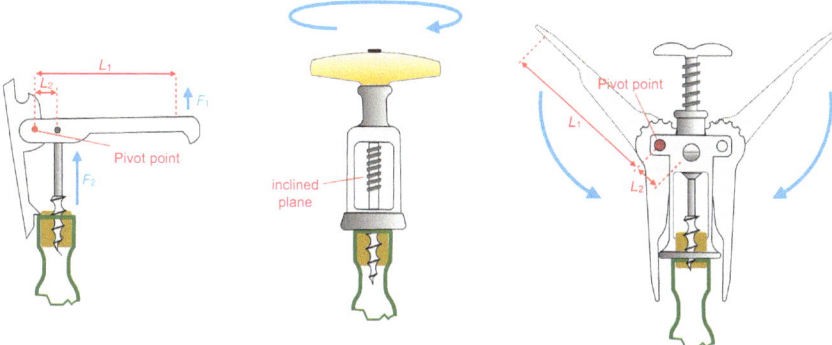

Fig. 1.15   Examples of different mechanisms of corkscrews

Fig. 1.16   Principle of the two-sided lever

Experiment 8: The Corkscrew as a Load Crane

Loads can be lifted effortlessly with the "Corkscrew crane". In a first sub-experiment with a spindle corkscrew (Fig. 1.15, middle) the cross handle is set so that it is "switched" to the thread. By slightly turning the cross handle, an attached mass piece is effortlessly lifted by a few centimeters.

In a second sub-experiment on a wing corkscrew, the forces acting on both lever arms and on the corkscrew spiral are determined with three spring force meters (Fig. 1.18). At the same time, the paths on the lever arms and the "stroke path" can be easily determined. The ratio of this lever path to the stroke path corresponds in good approximation to the ratio of

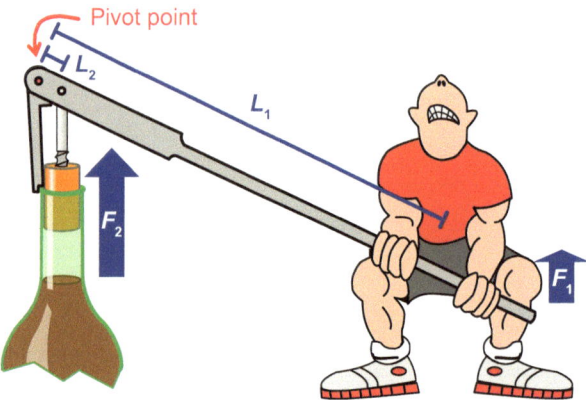

Fig. 1.17   Principle of the one-sided lever

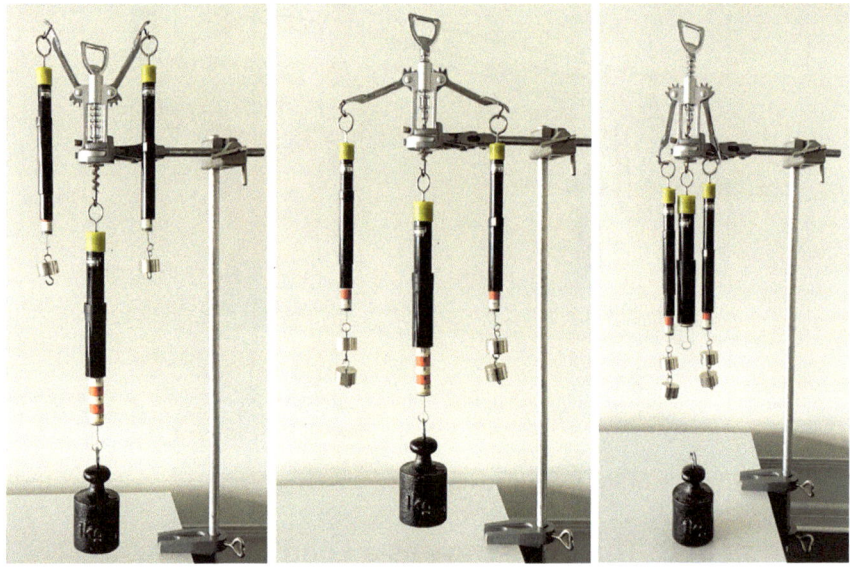

Fig. 1.18   Determination of the ratio "pulling force" to "lever force" on a wing cork-screw (left: the weight force on the two wings of 0.5 N each is not sufficient to lift the 1-kg piece. Middle: Here there is a force equilibrium between the two wings with 1.0 N each and the spiral with 10 N)

"pulling force" and "lever arm force" and is about 5:1 for the corkscrew used here. This ratio can then also be applied to the "real case", namely pulling out a cork with a force of only about 55 N with a wing corkscrew compared to the approximately 275 N with a simple corkscrew.

2

Well Ventilated? The Art of Decanting

This chapter is about the correct relationship of wine to air. It should breathe! Often, but not always, it is good for the wine to expose it to more air, or more precisely, to more oxygen, before enjoying it. Caution is particularly necessary with older and well-matured wines. The process of aerating wine is referred to by experts as "decanting". The standard variant involves slowly (!) pouring the wine from the bottle into a decanter and possibly letting it stand for a while. We will introduce variants here that bring the wine into contact with as much oxygen as possible while saving a significant amount of time. But beware: purists may already be warned here and skip experiment 12 when reading. Since oxygen supply is not always the method of choice and particularly opened bottles with high-quality content should even be protected from too much air supply, we have also addressed this problem in an experiment.

Different Variants of Decanting

Experiment 9: Swirling—Old-Fashioned Gesture?

Do you know any party guests with the connoisseur's eye—how they casually let the freshly poured glass circle in front of them before they start enjoying it? Is this just an old-fashioned gesture or does it have its justification? After all, it's about extracting the best possible aroma from the wine through the right aeration. We will pursue this question here. And compared to the other variants of aerating wine that will be presented later, the swirling of the glass seems to be the most natural and also the least effortful, which the guests also have to take care of themselves.

First, however, a few thoughts on the effect of aerating wine. Oxygen contact accompanies and influences the wine throughout its production and storage process. The modern wine industry uses a sophisticated control of dosage and duration of oxygen contact for the design of taste and aroma, but also the color intensity of their products. A forced oxygen supply of the wine after opening the bottle can, but does not necessarily have to be advantageous (see info box in experiment 14) and depends very much on the type, age and maturity of the wine. It is well researched what effects oxygen can have on the taste, aroma and color properties of wine. If the wine is unsulfured, oxygen leads to an increase of acetaldehyde (Nickolaus, 2018, p. 61), which is not without problems for health and also not beneficial for the aroma. Also tannin-rich red wines do not necessarily benefit in their taste profile from abundant oxygen contact. The relationship of tannins and polyphenols in red wines is related to the property of so-called *astringency*, a "frictional feeling" in the mouth perceived as unpleasant. As a result of oxygen, this sensation can be weakened, the wine then tastes "softer" or even "creamier". In addition, oxygen can also reduce the bitterness of the wine (Nickolaus, 2018, p. 61). Too much oxygen, on the other hand, can lead to increased browning and oxidative aromas, thus leading to taste defects in the wine.

So let's assume you know what you're doing, i.e., you know that the wine in your glass needs oxygen. How can you now also be sure that the swirling of the glass brings a significant input of oxygen to the wine?

Here we can help and provide an indirect argument with a small experiment. From chemistry, a nice standard demonstration experiment is known, which bears the name *Blue-Bottle-Experiment* (see info box). Behind it is a redox reaction in which an initially colorless solution is oxidized by atmospheric oxygen. However, in the classic execution, the solution must be shaken in a closed vessel so that the oxygen can diffuse into the solution. If you then let the solution stand—one minute is completely sufficient, a reduction takes place and the solution decolorizes again. The experiment is described in detail in Brandl (2006).

In a variation of this experiment, we wondered whether perhaps the mere swirling of the solution in a wine glass could cause the oxidation reaction and thus the blue coloration. Indeed, it works and Fig. 2.1 shows the clearly recognizable result.

The glass was filled with the methylene blue solution as one would do with wine. After it has settled, it takes on the (almost) decolored state

Fig. 2.1 Variant of the "Blue-Bottle-Reaction" in the swirled wine glass; before the swirling (**a**), after the swirling (**b**)

(Fig. 2.1a). Then we swirled the glass for about half a minute, during which the clearly recognizable blue coloration occurred (Fig. 2.1b). This shows that the process of swirling introduces so much additional oxygen into the swirled liquid that it leads to chemical reactions that would not occur without swirling.

"Blue-Bottle-Reaction"

The solution contains sodium hydroxide (NaOH) and glucose ($C_6H_{12}O_6$), which are dissolved in water. This solution is added to a round-bottom flask (or another sealable vessel, Fig. 2.2) with methylene blue solution ($C_{16}H_{18}N_3SCl$) and the vessel is sealed. Shaking the round-bottom flask causes the solution to turn blue. If the solution is left to stand quietly, it decolorizes again.

The Blue-Bottle-Reaction represents a redox system in which the methylene blue serves as a redox indicator. The decolorization is based on the reduction of the dye methylene blue to colorless leukomethylene blue by the alkaline glucose solution. In this process, the glucose is oxidized to gluconic acid ($C_6H_{12}O_7$).

If you shake the solution, oxygen enters the solution and the leukomethylene blue is reoxidized to colored methylene blueMethylene Blue oxidized. The experiment can be repeated as long as there is glucose in the solution to reduce the methylene blue or the available oxygen is consumed.

Methylene blue is used in the textile industry for dyeing fibers and in the paper industry also as printing ink. In medicine and microscopy, it is used as a so-called vital dye.

Fig. 2.2 "Blue-Bottle-Reaction" in the sealed flask

In addition to aerating the wine and better development of its aromas, swirling also allows for a mechanical observation. The movement creates wave crests and troughs that circle in the glass and cover the glass wall. Sometimes only one large wave crest is visible, but sometimes several smaller ones. Reclari et al. (2014) were able to show that the resulting wave formation depends on three factors. These are the ratio between the fill level H and the glass diameter D, the ratio between the diameter of the circular motion d and the glass diameter D as well as the ratio of the acting forces (centripetal force F_z to gravitational force F_g; Fig. 2.3):

$$\tilde{H} = \frac{H}{D} \quad \tilde{d} = \frac{d}{D} \quad \tilde{F} = \frac{F_z}{F_g} = \frac{\omega^2 d}{g}$$

(ω: angular velocity; g: acceleration due to gravity)

By slightly varying these parameters—you will not want to keep the level of your glass constant all evening anyway—you can therefore create very different wave patterns in your glass (Fig. 2.4).

Reclari et al. also found that completely independent of the size of two vessels, exactly the same wave formation occurs, as long as only the ratios introduced above match each other. The observations they made in the wine glass and on rotating cylinders of very different sizes can therefore be quite interesting for the industry, for example when mixing liquids.

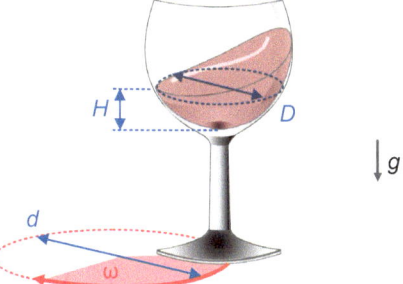

Fig. 2.3 Physical quantities determining the wave formation in the wine glass

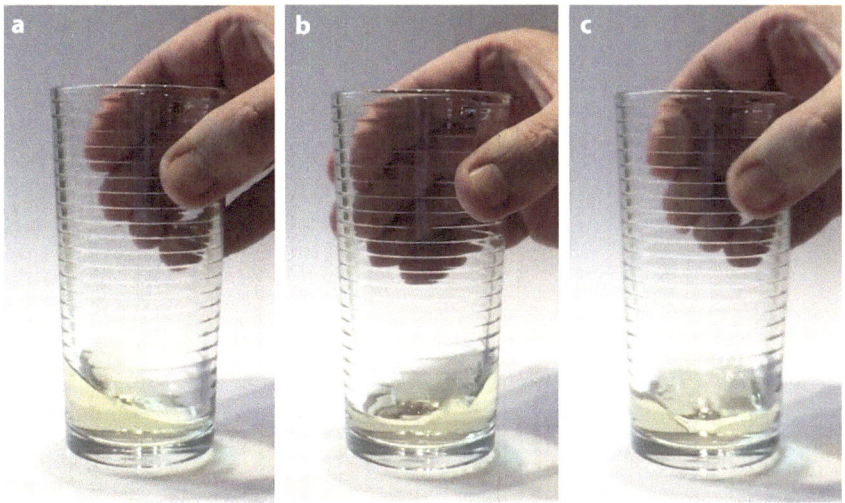

Fig. 2.4 Different number of wave peaks at the same fill level due to variation of the angular velocity; only one wave peak (**a**), before the large wave peak on the right a smaller one can be clearly seen (**b**), at least four wave peaks (c)

Experiment 10: The Surface Counts—What Lasts Long …

If you want to spare your guests the swirling and present the wine already well aerated, the mentioned decanting using a decanter is the usual proce- dure. For decanting, i.e. to separate the wine from its depot (formed sed- iment and tartar), no special carafe is needed. However, for aeration or decanting, it should definitely be a carafe with as wide a base as possible (Fig. 2.5a). The large radius leads to a large contact surface between wine and air, which can significantly speed up the aeration process (usually 1–3 h). You can easily test this with a simple experiment: Pour the same amount of a suitable red wine for decanting into a decanting carafe and into a cylindrical water glass. Let the wine breathe for about 3 h and compare the taste. You will probably find that the wine from the decanter appears softer, more aromatic and possibly also less bitter. This would already prove the effectiveness of aeration using a decanting carafe! By the way, you can also speed up the process by occasionally swirling the decanter.

However, since the taste experiment cannot be considered as objective evidence, we have increased the decanting duration from 3 h to several days. The wine is then heavily oxidized and no longer enjoyable, so you should refrain from this extension and trust our result. This is shown in Fig. 2.6 and now also objectively demonstrates the effectiveness of the decanting carafe. The wine aerated in the decanter has discolored much more than the one from the water glass. The only cause for this can be a stronger oxidation, which in turn results from the larger contact surface with oxygen. In our experiment, the carafe has a radius of about 9 cm, that of the water glass is only 3 cm. Therefore, the contact surface with the air is 9 times larger in the decanter!

Fig. 2.5 The same amount of wine in a decanter (**a**) and a water glass (**b**) immediately after opening the bottle

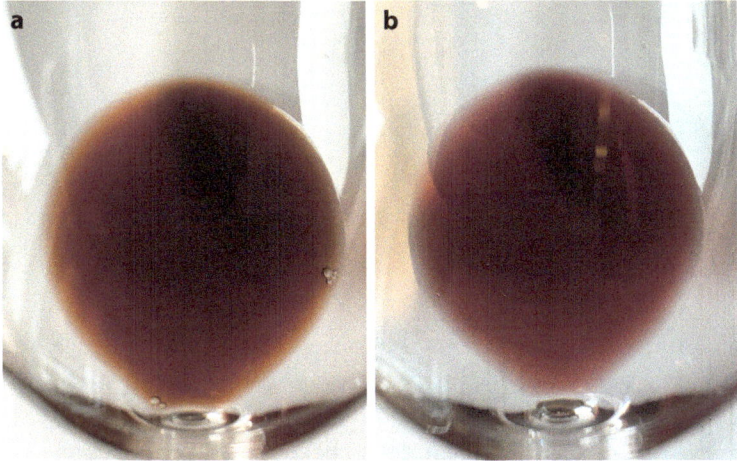

Fig. 2.6 Wine from decanter (**a**) and water glass (**b**), several days after opening the bottle

Experiment 11: Clever! Primitivo a la Venturi

This method of aeration is incredibly time-saving. The pouring itself already leads to enhanced wine aeration. You simply put a special pourer on the opened wine bottle and don't have to worry about the oxygen supply anymore. This is taken care of by the so-called *Venturi nozzle* built into the pourer. Its decisive feature can be seen on closer inspection: the wine has to flow through a constriction when pouring, at the narrowest point of which there is a small opening to the outside (Fig. 2.7).

To understand the Venturi pourer, we must first accept a self-evident fact: The amount that flows into the pourer from the back of the wine bottle comes out again at the front end. Otherwise, wine would have to accumulate in front of the pourer or wine would be lost behind the pourer. Both would be very strange.

From this assumption and the further assumption that the density of the wine does not change when pouring, a law emerges, which is referred to as the *law of continuity* (see info box).

The associated continuity equation states that at the location of a smaller cross-sectional area in a liquid line, the flow velocity is greater than at a location with a larger cross-sectional area.

This is exactly what is exploited in this wine pourer. The narrowing in the pourer thus ensures a high flow velocity. This in turn leads us to another law, namely the relationship between pressure and flow velocity: The pressure in

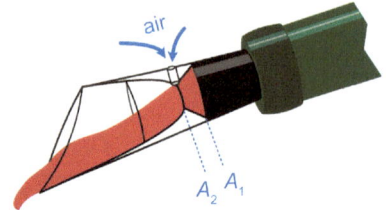

Fig. 2.7 Venturi pourer in action

a flowing liquid is the smaller, the greater the flow velocity at this point. Quantitatively, this fact is described by the *Bernoulli equation*. In this way, it is possible that as a result of the high flow velocity at the narrowing, the pressure is reduced and ambient air is sucked in through the opening introduced exactly at the narrowest point (Fig. 2.7). The sucked in air ensures pressure compensation and "pearls" into the flowing wine. This achieves the goal of a better aeration.

Law of Continuity and Bernoulli Equation

For an incompressible liquid flowing through a pipe at two points of different cross-sections A_1 and A_2 with the velocities v_1 and v_2, the following applies:

$$A_1 \cdot v_1 = A_2 \cdot v_2 \quad \text{resp.} \quad \frac{A_1}{A_2} = \frac{v_2}{v_1}$$

The ratio of two flow cross-sections is equal to the inverse ratio of the flow velocities associated with these cross-sections. This is referred to as the *law of continuity*.

When a flowing medium transitions from a point of larger cross-section to a narrow point, the medium thus gains kinetic energy. Due to the conservation of energy, this increase occurs at the expense of the volume work performed in the pipe. From this consideration, for a horizontal flow (without the influence of gravity), the *Bernoulli equation* results:

$$p + \frac{1}{2}\rho v^2 = \text{constant}$$

p is the *static pressure*, which a pressure probe measures for a tangentially flowing medium. The expression $\frac{1}{2}\rho v^2$ with the density ρ of the medium indicates the *stagnation pressure*.

From the Bernoulli equation, it follows that the static pressure in a flowing medium is the lower, the greater the flow velocity becomes.

A construction, as used in the wine pourer described here, was first developed in 1797 by *G. B. Venturi* and is therefore also called a *Venturi nozzle*. It is probably safe to assume that Venturi did not first think of wine aeration when he made his invention. Other applications of this physical principle are more familiar from everyday life or school. For example, in a Bunsen burner, the air oxygen required for the combustion of the gas is sucked in as a result of the combustion gas flowing through a constriction similar to that of the wine pourer.

As convenient as the Venturi pourer is, the taste improvement achieved with it for red wine is not yet completely convincing. A nearly equally fast, but significantly more effective wine aeration will be described in the next experiment. But beware: This is not for the faint-hearted!

Experiment 12: For Those in a Hurry—Flash Aeration with the Smoothie Maker

With this variant, we might possibly drive a trained sommelier to the brink of a nervous breakdown, and at the very least it would be an challenge. If you also like to let your wine breathe slowly, i.e. 1 to 3 h, you can safely skip this experiment. If you find the swirling of wine too pretentious, this method is probably just right for you. It's about the so-called hyperdecanting, first described by *Nathan Myhrvold, Chris Young* and *Maxime Bilet* in their multi-volume cookbook *Modernist Cuisine*. The procedure is extremely simple: You put the wine into a kitchen blender or a smoothie maker, "beat" it for about 40 s, and the aeration is done (Fig. 2.8)! For a nicer presentation of your good drop, you can then pour it into a decanter or back into the original bottle using a funnel.

At this point, we would like to invite you not to a physical experiment, but to a taste experiment, i.e. to a blind tasting. For this, you will need, in addition to the blender or smoothie maker, another person and a wine suitable for decanting. We would like to recommend a very simple and inexpensive drop, e.g. a very young red wine from the discount store in the price range of 3 to 5 US$. With such a wine, aeration can work wonders, which will benefit you in the blind tasting—you will be thrilled by the effect!

So that no cheating can be accused of you, you first leave the scene. Your co-experimenter pours the wine into the kitchen machine, filling the vessel only up to a third at most. Since at least a sip of the unaerated wine is needed for the tasting, the bottle must not be completely emptied. After an aeration time of about 40 s, the experimenter pours the same amounts

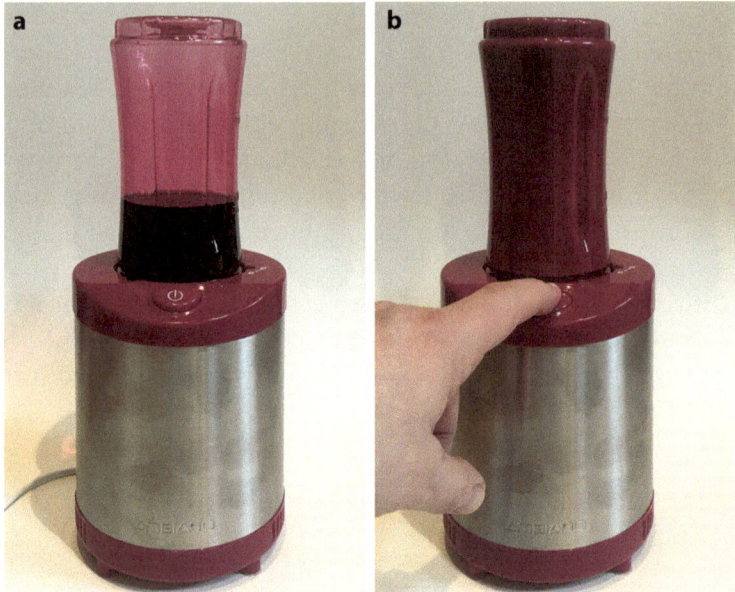

Fig. 2.8 Aerating wine with a smoothie maker

from the bottle or from the mixer into two identical glasses and invites you in. The tasting can begin and we assure you that the wine from the mixer will taste significantly softer and more aromatic—you will easily identify it correctly.

The experiment only works with initially unaerated wine, i.e., the bottle must only be opened immediately before the blind tasting and not, for example, the day before.

Why does this turbo aeration in the mixer or smoothie maker work so well? As we have already seen with decanting using a decanter, the contact surface between wine and air is the decisive factor. The larger this contact surface is, the more oxygen can get into the wine in the same time—the aromas develop faster and the tannins are softened more quickly. The strong swirling of the wine maximizes its surface and thus it receives a concentrated oxygen supply in the shortest possible time.

However, you should not overdo it with the mixing, as you also beat aromas out of the wine. This can be observed when opening the mixer, where an aroma cloud is literally released. With the recommended simple wine and an aeration time of 40 s, our experience shows that the positive effects clearly outweigh!

For Those in a Hurry: Turbo Pouring with Pressure and Tornado

Experiment 13: Pouring Time Halved Twice!

Now that the wine has already sufficiently "breathed" and we have also seen a solution for those in a hurry, let's stay a little longer with the, let's call it, efficient approach. If the aeration with the smoothie maker has already been a challenge for wine aesthetes, then those people might want to skip the following experiment. This also does not quite correspond to the usual table manners. However, there could be life situations that require decisive action—in this case, pouring as quickly as possible. If one can engage with this, then the question to answer would be: How quickly can a bottle of wine actually be emptied?

Let's try this for practice purposes with water-filled wine bottles and start with pouring from the bottle simply held upside down (Fig. 2.9a). You hear the familiar "glugging", which is caused by water and air having to squeeze past each other in the narrow bottle neck—water out and air in. This happens through a rhythmic sequence of air bubbles rising and smaller amounts of water falling more or less downwards. The liquid literally struggles out of the bottle. The manually timed duration for this emptying method is about 8 s – too long for those in a hurry. Also, systematically varying the "pouring angle" does not yield shorter times.

To beat this time, a hole in the bottom of the bottle would be helpful. Then the air could go in there and the liquid could flow out of the bottle neck undisturbed. However, a hole is difficult to make in a glass bottle. Nevertheless, this consideration leads in the right direction. An extra hole for the air is needed!

We create this hole not in the glass material, but as a "hole in the hole" through the bottle neck. For this, we use the following trick: The bottle is set in circular motion at the beginning of the pouring (Fig. 2.9b).

As a result, the liquid in the bottle also begins to rotate due to friction processes and clings to the outer glass wall. From the upper level in the bottle to the outflow opening in the bottle neck, a parabolic vortex funnel forms, which works its way all the way down through the bottle neck. The necessary air hole is thus created! Once the emptying vortex has formed, it usually remains until the bottle is completely emptied. And this is significantly faster than just tipping the bottle over. In smaller wineries, where wine bottles are hand-washed in large vats, the time saved by this emptying

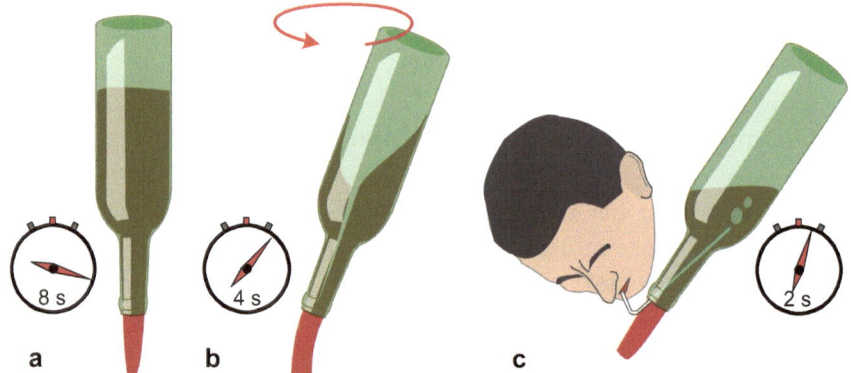

Fig. 2.9 "High-speed techniques" for emptying a bottle

method is actually an economic factor. For the same bottle, a time of about 4 s is clocked with the "tornado method". That's a record! Or is it?

Well, it can actually be done even faster. However, the potential guests at the table then have to be even more tolerant. The person pouring takes a bent drinking straw into their mouth at the short end. The long end is inserted into the bottle, the bottle is turned upside down and the pourer blows hard into the straw, while the bottle content now literally shoots out. In 2 s the bottle is empty! That's really a record—although not quite socially acceptable anymore.

When the Wine Should not Breathe …

So far, we have been talking about aerating the wine after opening the bottle. As a rule, this results in a noticeable improvement for good quality red wines. However, it can certainly happen that a bottle is not emptied in one evening. That's no shame! Then a residual amount remains in the bottle. This remaining wine continues to "breathe", even if the bottle has been resealed with a cork or otherwise. Because there is air above the wine, it is generally true that the wine should be enjoyed as soon as possible after opening.

In addition, there are also wines for which aeration is not beneficial from the outset. This mainly affects long-matured treasures. Here, possibly even decanting—i.e. transferring—leads to too intense contact with oxygen. The rarity can lose its aroma within minutes. For the enjoyment of such noble

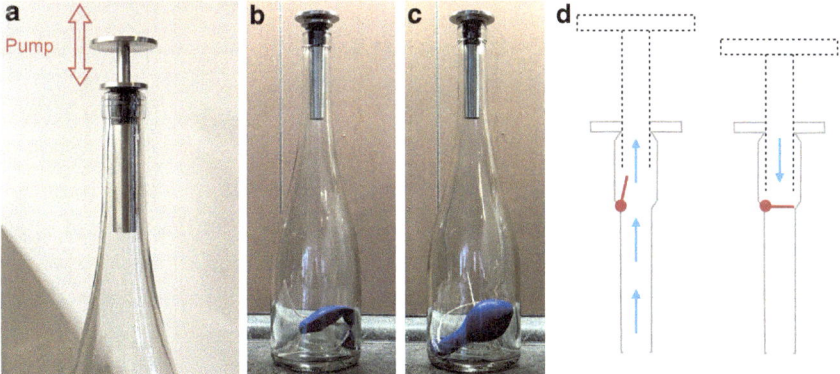

Fig. 2.10 Evidence of the negative pressure created by a vacuum wine pump in a wine bottle; commercial vacuum wine pump (**a**); sealed balloon with little filling in the bottle before pumping (**b**); inflated balloon after pumping (**c**); functional diagram of the pump with backflow valve (**d**)

wines, it is best to invite as many good friends as possible so that no residue remains in the bottle.

Manufacurers have given us a small tool for such cases, which we will examine in the following: the vacuum wine pump. It is available in various designs and price ranges. An example is shown in Fig. 2.10a.

When *Otto von Guericke* invented the piston vacuum pump in the 17th century, it had significant effects on the further development of physics and even philosophy. Whether the introduction of the vacuum pump into the household will have a similarly dramatic effect on wine culture is for you as connoisseurs to judge. In any case, we can take a physical look—supported by an experiment—at the corresponding tool.

Experiment 14: Balloon and Vacuum Wine Pump

First, we should briefly clarify the term "vacuum". In its pure form, this state describes a space that contains no or so few molecules that they do not interact with each other, but only with the vessel wall. In technology and physics, the term is differentiated: At a gas pressure between 1000 mbar (just below atmospheric normal pressure) and 1 mbar, it is called a *rough vacuum*. Between 1 mbar and 10^{-3} mbar it is a *fine vacuum* and below a pressure of 10^{-3} mbar, the *high vacuum* is reached (Gerthsen, 1999, p. 279). We can therefore assume that with our vacuum pump, according to this categorization, we will at best end up in a rough vacuum. This also applies to all "vacuum" packaging often found in households and everyday life, such as for coffee or peanuts.

But can we determine, or show, that the pump actually creates a negative pressure? If it is possible to place a second, smaller and closed volume with flexible vessel walls in the wine bottle, then a decrease in pressure in the bottle could be recognized by observing the behavior of this body when evacuating. A balloon is of course predestined for such a flexible and sealable body. But anyone who has ever tried to inflate a balloon in a bottle will have encountered an insurmountable resistance. As soon as the balloon reaches the circumference of the bottle or even the bottle neck, it blocks the bottle exit for the air still contained in the bottle. Nothing works anymore!

But it is not necessary to inflate the balloon properly. It is enough to equip it with a very small amount of air. Even then, it will still be difficult to get it inside through the bottle neck. A straw provides a solution, offering the air from inside the bottle an escape route to the outside. Fig. 2.10b shows the almost limp balloon in the bottle. Now the vacuum wine pump is put on and pumped properly. It takes many strokes until the result becomes visible. But finally, the little remaining air in the balloon starts to show. The limp shell inflates—a clear sign that the pressure inside the bottle has become lower (Fig. 2.10c).

The operating principle of the wine pump is based on the principle of a *backflow valve* (Fig. 2.10d). When the air is pulled out, the check valve opens. Due to the then greater pressure outside the bottle, the flap is pressed firmly until no more air is pumped out.

As a conclusion of our small experiment, we summarize: The wine vacuum pump we used does what it is supposed to do. It pumps a detectable part of the air out of the bottle. Thus, we can give a purchase recommendation for single households or for preferences for well-aged red wines.

Unstable Wine in Contact with Air

Wine or the mash is exposed to atmospheric oxygen and various bacteria throughout the entire production process. Through oxidation and metabolic processes of microorganisms, the aroma or color properties of the wine are constantly influenced.

Oxidation or fermentation. The acetic acid bacteria contained in the air cause the alcohol (ethanol) to be broken down into acetic acid (ethanoic acid). The reaction that takes place is as follows:

$$C_2H_6O + O_2 \longrightarrow C_2H_4O_2 + H_2O$$

Acetaldehyde (ethanal). This carcinogenic substance is formed in the human body during the breakdown of alcohol and leads, among other things, to the "hangover symptoms", but can also lead to irreversible organ damage. Acetaldehyde is also already present in the wine, where it is produced during fermentation from the metabolism of yeast strains and bacteria. In addition, it is also produced by storage with air contact (e.g. in leaky barrels) in the fermented wine. Part of the oxygen reacts with the ethanol to form acetaldehyde.

Change of color properties. The plant pigments that give red wine its color are anthocyanins (see also Experiment 25 "Red Wine as Color Filter"). Through oxidation reactions, these pigments can be broken down. The degradation products often have a brownish color, so the intense red tones of a wine tend to brown increasingly when stored with air supply. Anthocyanins are also destroyed by light, which is why a longer bottle storage should definitely be protected from light.

Sulfur dioxide in wine—why? "Contains sulfites" is a note to be read on almost all wine labels. Although ethanol itself is a "natural" preservative of wine, the alcohol content is not sufficient to prevent the unwanted metabolic effects of microorganisms. Here, the addition of sulfur dioxide helps, as it acts both antimicrobial and antioxidant, thus contributing to a significantly better storage ability of the wines.

In wine, sulfurous acid (H_2SO_3) is present in three forms:

1. as undissociated sulfur dioxide (SO_2)
2. dissociated: $SO_2 + H_2O \longrightarrow H^+ + HSO_3^-$
3. in sulfite form: SO_3^{2-} (sulfite anion)

With the addition of up to 30 mg/l sulfite (depending on the quality of the harvested wine berries or the sugar content of the wine), not only are oxidations prevented, but also unwanted secondary fermentations, e.g. in wines with high residual sweetness.

Even when sulfur dioxide is added, it is always the case that "natural" sulfur dioxide is already contained in the wine, which is produced during the alcoholic fermentation as a metabolic product of the yeast cells.

3

The Ear Drinks Too: Acoustics with Wine Glasses and Bottles

Now we are well prepared: The bottles are open, the wine has breathed properly, and we finally allow ourselves a first taste! And of course, this is ringed in by the beautiful ritual of toasting. "Ring in" seems particularly apt here, as it brings the clinking of glasses into the conceptual proximity of the ringing of bells. We will return to this—also acoustically fitting—analogy. Why glasses ring at all and how their sounds can be measured with simple arrangements is the subject of a series of experiments. In the further course, it is not only about stimulating our readers. We also stimulate glasses! The so-called resonance frequencies are associated with the individual sounds of such stimulated glasses. We will use these for a simple determination of the speed of sound. Finally, the acoustic properties of bottles and glasses lead us to very special musical instruments.

Acoustics with Wine Glasses

Vibrating bodies, such as strings, membranes or sound bars, cause pressure fluctuations in their immediate surroundings, which spread through the room in the form of sound waves. The situation is completely analogous with a struck wine glass, which also vibrates at its characteristic natural frequency and produces a corresponding sound. Its frequency spectrum provides an "acoustic fingerprint" that we would like to examine more closely before we come to the clinking of two glasses.

L. Kasper and P. Vogt, *Uncorking the Physics of Wine*, https://doi.org/10.1007/978-3-662-68759-8_3

Experiment 15: Frequency of Wine Glasses

To analyze the sound of a vibrating red wine glass, we lightly strike it with a wooden spoon and analyze the resulting sound with a suitable smartphone app (e.g., Sound Analyzer, Fig. 3.1). It turns out that the sound signal produced by the wine glass is created by the superposition of individual tones. Even though we would speak of the "tone" of the glass in everyday life, it should—physically correct—actually be called "sound" (see info box). The first spectral line in the frequency spectrum is at 581 Hz, it corresponds to the fundamental frequency f_0 of the sound. The other spectral lines mark the so-called overtones with the frequencies measured in the experiment $f_1 = 1487$ Hz, $f_2 = 2745$ Hz and $f_3 = 4038$ Hz. The surprising thing about this result is that with musical instruments with one-dimensional sound generators—e.g., vibrating strings (chordophones) or vibrating air columns (aerophones)—the upper frequencies are always integer multiples of the fundamental frequency (Fig. 3.4b; Table 3.1). With the used wine glass, which has a bell-like shape, however, we observe the frequency ratios 2.6, 4.7, and 7.0 to the fundamental tone (Fig. 3.2). Our measurement deviates

Fig. 3.1 Sound analysis of a struck red wine glass

Table 3.1 Frequencies of the fundamental tone and the overtones of an A4 played on the piano

	Measured frequency in Hz	Frequency ratio to the fundamental tone
Fundamental tone	441	1
1st overtone	872	$1.98 \approx 2$
2nd overtone	1314	$2.98 \approx 3$
3rd overtone	1755	$3.98 \approx 4$
4th overtone	2208	$5.01 \approx 5$
5th overtone	2660	$6.03 \approx 6$
6th overtone	3112	$7.06 \approx 7$

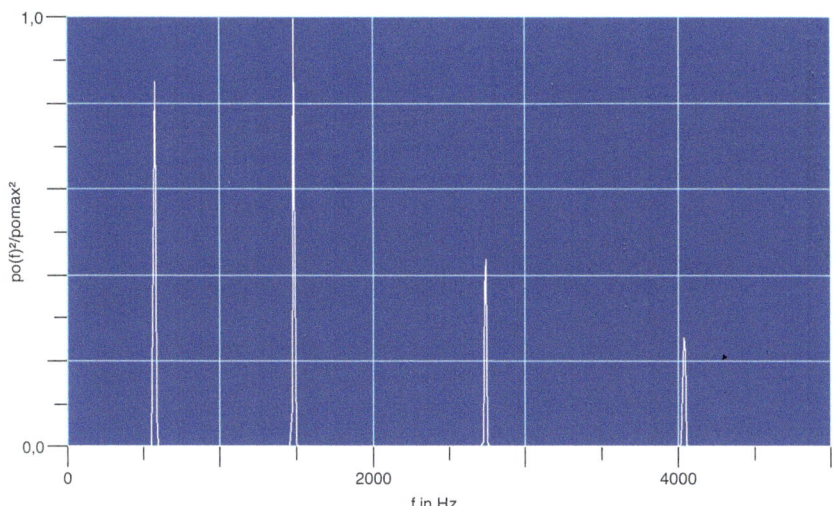

Fig. 3.2 Frequency spectrum of a struck red wine glass

somewhat from the theory, according to Denninger (2013) one would actually expect the ratios 2.3, 4.0, and 6.25, but it is actually not so bad.

Even if you do not play music yourself, you will surely find a recorder, or a children wind harmonica at home. So try it out and determine the frequencies of the fundamental and overtones of existing musical instruments and wine glasses. You can also experiment with other objects, e.g. with singing bowls and porcelain bowls, or you analyze the sound of the bell of a nearby church (see info box).

Compared to vibrating strings or air columns, the calculation of the fundamental frequency of a vibrating glass is significantly more complex. This depends on the speed of sound in glass ($v_s = 5300$ m/s), the glass thickness d (in our example this was 1.16 mm) and the geometry of the glass.

For a cylindrical glass, the fundamental frequency is calculated according to Schlichting and Ucke (1995) as follows:

$$f_0 = \frac{v_s \cdot d}{\sqrt{3}\pi R^2}$$

For an estimate, we now apply this relationship to our red wine glass. Inserting the numerical values leads to 640 Hz for a mean radius R of the used glass of 42 mm and thus agrees well with the experimental result.

Types of sound

A *tone* always occurs when the vibration of the sound-emitting body can be described by a single sine function, i.e., when it is a harmonic vibration (Fig. 3.3a). If one subjects the acoustic signal of a tone to a Fourier analysis and presents the result in the form of a frequency spectrum, one obtains a single spectral line at the frequency f. A measurement example is shown in Fig. 3.3b.

If one records the sound signal of a musical instrument and presents the oscillogram graphically using an analysis app, a periodic, but usually not sinusoidal, vibration pattern results (Fig. 3.4a); one speaks of a *sound*. According to the *Fourier theorem*, such a signal can be represented as a sum of sine functions, whose arguments are integer multiples of a fundamental frequency (Table 3.1, for the FFT analysis see also info box for experiment 18). The frequency spectrum of a sound has, in contrast to that of a tone, several spectral lines (Fig. 3.4b). The perceived pitch of a sound is influenced solely by the fundamental frequency, the number and the amplitudes of the overtones determine (in addition to the transient processes) the timbre of the instrument. Through them, we can easily distinguish between two instruments, even if they play the same note at the same volume.

In contrast to the tone and sound, a *noise* (e.g., crumpling a sheet of paper) is not caused by periodic processes. The Fourier analysis provides an almost continuous spectrum (noise), i.e., there are a multitude of individual tones present, which can take on arbitrary frequencies.

We perceive a sudden mechanical vibration of large amplitude and short decay time as a *bang*. Examples of this are the bursting of a balloon or clapping hands. Similar to a noise, a bang cannot be assigned a single frequency, but only a frequency range.

Frequency of church bells

Church or carillon bells are musical instruments that can be found almost everywhere in everyday life and can be easily examined with a smartphone. They are very similar in shape and therefore also in their acoustic properties to wine glasses. Their physical background theory proves to be difficult and a reliable prediction of their natural frequencies is only possible from the exact dimensions using the finite element method. If you ask a bell founder how he

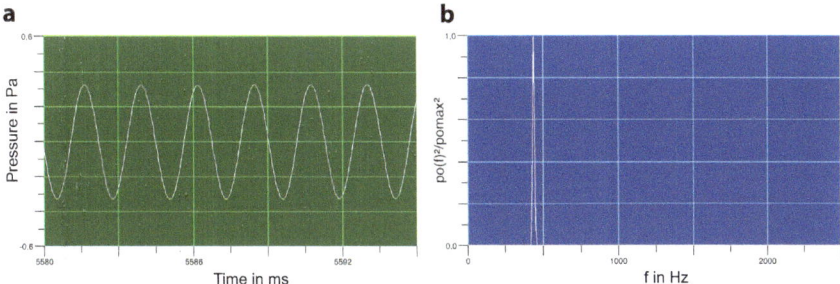

Fig. 3.3 Oscillogram (**a**) and frequency spectrum (**b**) of the tone of a tuning fork

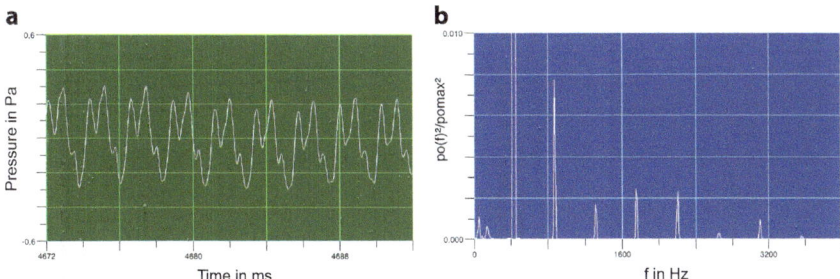

Fig. 3.4 Oscillogram (**a**) and frequency spectrum (**b**) of the sound of a piano (note A4)

calculates the rib (half longitudinal section of the bell) for a bell with a desired frequency spectrum, you will certainly not get any information: The art of bell casting is based on centuries of experience and the knowledge about rib construction is passed on exclusively to direct descendants. However, we would like to present the result of a modeling that could be further improved by comparing it with a dataset comprising almost 700 bells (Vogt et al., 2015, 2016) and that cannot be surpassed in simplicity! The frequency of the fundamental tone f_0 of a church bell is:

$$f_0 = \frac{100 \text{ m/s}}{R}$$

Or in words: 100 divided by the measured fundamental frequency of the bell in Hz gives a good approximation of the bell radius in meters. The average deviation of the estimate made with this rule of thumb from the actual bell radii is just 3.5%!

Experiment 16: Cheers! Beat frequencies when clinking glasses

After we have dealt more closely with the frequency of a vibrating wine glass, we now want to turn to the simultaneous vibration of two glasses. You are well familiar with the experimental situation from clinking glasses together, but have you ever paid close attention to the sound that is produced? At least when the glasses are filled to similar levels, you hear a tone whose volume varies periodically—it constantly gets louder and quieter (from a physical point of view, "tone" is strictly speaking a wrong term at this point, but that should not bother us further, see info box of the preceding experiment). This phenomenon is called *acoustic beat* and the number of volume changes per second corresponds in absolute terms to the difference between the two initial tones (beat frequency). By the way, two wine glasses of the same model are never completely identical and therefore have slightly different frequencies. Consequently, the wine glasses do not even need to be filled to produce an acoustic beat. You can observe a beat even when they are empty, which admittedly would be equivalent to a sad toast.

With the help of a smartphone and a suitable sound analysis app (we used the *Spaichinger Sound Analyzer* for this), the beat can be visualized and even quantitatively evaluated (Fig. 3.5; Vogt & Kasper, 2021b).

For the quantitative evaluation of the experiment, the frequencies of the glasses used are first determined one after the other and these are lightly struck, for example, with a wooden spoon. In the experimental example, these were 581 Hz and 592 Hz respectively (Fig. 3.6). Taking into account the law formulated in the info box, one would therefore expect that the superimposed signal reaches a volume maximum 11 times in 1 s.

To verify, an oscillogram of the overlay can be displayed. As can be seen in Fig. 3.7, the acoustic signal reaches a volume maximum 17 times in 1.34 s, which corresponds to 12.7 volume changes per second. Considering that only the simplest arrangement was used for the experiment, the comparison with the theoretical value provides a satisfactory result.

Acoustic Beat

A special form of superposition of sound waves is the acoustic beat. It always occurs when at least two vibrations with a small frequency difference overlap. The observable auditory impression then corresponds to a tone, whose volume varies periodically. If the amplitudes of the initial tones are equal, the volume

Fig. 3.5 Quantitative analysis of the acoustic beat produced by two wine glasses

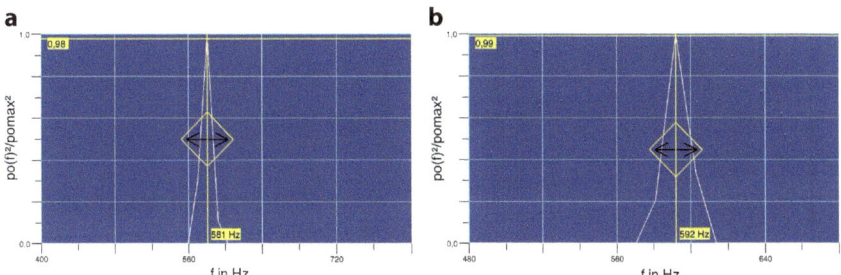

Fig. 3.6 Frequency spectra of the individual tones produced by the wine glasses

between the maxima drops to zero (*complete beat,* Fig. 3.8a), with unequal amplitudes, there is a so-called *incomplete beat* (Fig. 3.8b).

The number of volume changes per second is referred to as beat frequency f_s, which depends on the initial frequencies f_1 and f_2 and corresponds to the absolute value of their difference. So it follows:

$$f_s = |f_1 - f_2|$$

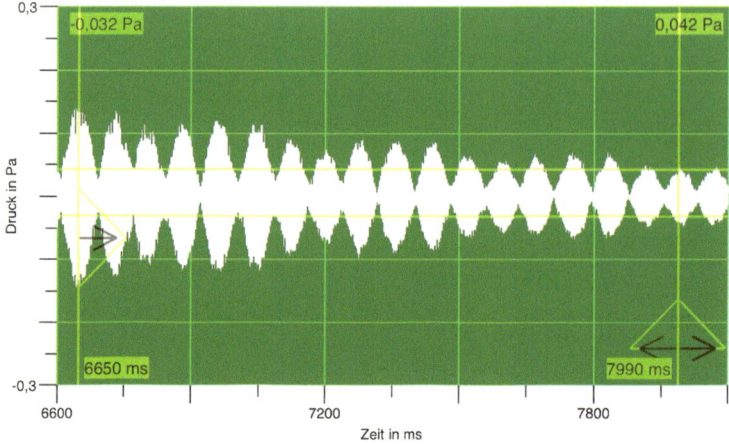

Fig. 3.7 Determination of the beat frequency

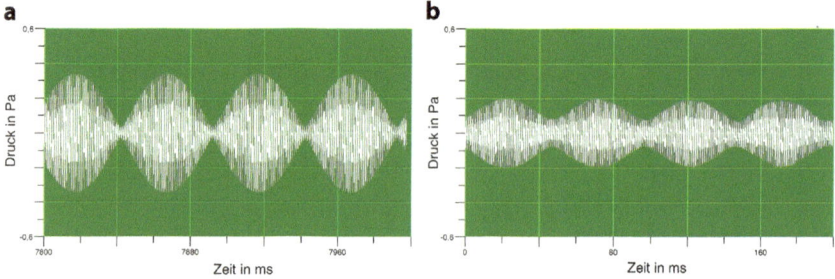

Fig. 3.8 Oscillograms of two beats; complete beat (**a**), incomplete beat (**b**), analyzed and displayed with the "Sound Analyzer" app

The frequency of the audible tone corresponds to the average of the initial frequencies, which, however, should not be considered in the experiment described here.

Tuning of Musical Instruments

Today, musical instruments are usually tuned with electronic tuning devices. In the past, tuning forks were used for this purpose, which produced a tone of known frequency (usually an A4 with 443 Hz). The phenomenon of acoustic beat played a crucial role in tuning musical instruments using a tuning fork. For example, if the a-string of a violin was out of tune by 7 Hz, a beat frequency of 7 Hz would occur. The string tension was then gradually varied until no beat occurred. The remaining strings were then tuned starting from the a-string.

This procedure is still used today in some cases, for example, when a violinist plays together with a pianist. The piano is not easily tunable, which is why it is used as a starting point for tuning the violin.

Experiment 17: Shattering Wine Glasses?—Recording a Resonance Curve

It is often claimed that people with trained voices are capable of shattering glasses by singing, and numerous videos circulating on the internet would have us believe this as well. The physical explanation often read is quite plausible: One would simply have to accoustically stimulate the glass with its natural frequency, causing it to vibrate particularly strongly. If one holds the tone long enough, the vibration would continue to build up and it would lead to a resonance catastrophe.

Indeed, the destruction of a glass in the described manner is possible when it is exposed to a tone generator. Unlike the human voice, this is capable of maintaining the frequency of the generated tone for a longer period of time. With the help of an amplifier, a significantly higher sound pressure can also be achieved than would be possible with the voice. And indeed, as numerous studies have shown, the desired resonance catastrophe requires a sound pressure that exceeds that of the voice by a hundredfold.

We now want to conduct the experiment on a small scale and analyze the phenomenon of resonance, that is, the excitation of a vibration with its natural frequency (see info box), in more detail. Glasses will definitely not break in the process!

To conduct our experiment, we first need to determine the natural frequency of the glass used in a preliminary experiment. We have already seen how we can estimate and experimentally determine this in Experiment 15. We strike the glass with a wooden spoon (Fig. 3.1) and display the frequency spectrum of the resulting sound signal with a suitable app (Fig. 3.9). For the glass used here, we find a fundamental tone of 603 Hz, which corresponds to the first natural frequency of the glass. If you want to excite the wine glass to vibrate solely by sound waves, this frequency should work best.

To verify this, we use a smartphone or a tablet computer with a tone generator app. We used the "Audio Kit" app for this, which allows us to generate tones accurate to 1 Hz. We expose the wine glass to this for about 7 s from a few centimeters away—this should be enough to excite a vibration -, turn off the tone generator and measure the sound pressure level of the tone still generated by the wine glass (Fig. 3.10). We used the free "phyphox" app for this, which can also be used for tone generation. We apply the described procedure for the natural frequency of 603 Hz as well as for the integer frequencies in the range of 593 Hz to 614 Hz. A measurement example is shown in Fig. 3.11. The course of the amplitude of the entire process is

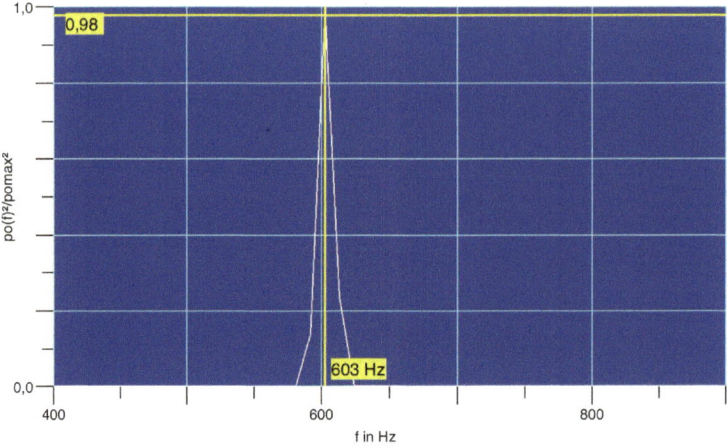

Fig. 3.9 Experimental determination of the natural frequency of the glass used

Fig. 3.10 Experimental setup for recording a resonance curve

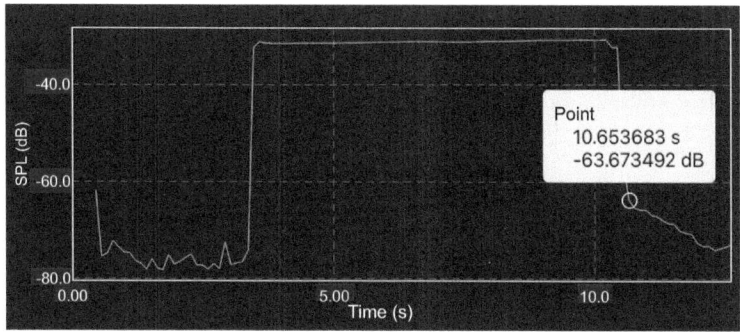

Fig. 3.11   Time course of the amplitude with a stimulation at 614 Hz

shown. The strong increase in volume marks the switching on of the tone generator app, which generated a tone of constant amplitude for about 7 s. After switching off the tone, the curve immediately drops to the sound pressure level caused by the glass. This value is read off and serves us as a measure of the excitation of the vibration (Fig. 3.11). The fact that the sound pressure level here is negative at around −63.7 dB does not need to worry us. Above the threshold of hearing, one would actually expect exclusively positive level values, at least if the app was calibrated correctly. However, we have dispensed with this calibration, as the absolute levels are irrelevant for the experiment.

The result of the entire series of measurements is shown in Fig. 3.12. From the resonance curve depicted there, we can deduce that the excitation of the wine glass at the natural frequency (which is apparently 604 Hz and not 603 Hz) works particularly well. The further we move away from the natural frequency, the quieter the wine glass resonates after the tone generator app is switched off. If we expose the wine glass to a frequency that deviates by at least 20 Hz from the natural frequency, no noticeable vibration excitation occurs anymore.

The experiment shows that to shatter the glass with singing, we would really have to hit exactly its natural frequency. Only then can the glass be excited to strong vibrations and be destroyed. However, due to our anatomy, we are not able to maintain a frequency exactly over a longer period of time. And even if we could, the sound pressure would be far from sufficient. Even a pitch-perfect opera singer can slightly vibrate the wine glass with her voice, but she cannot possibly destroy it.

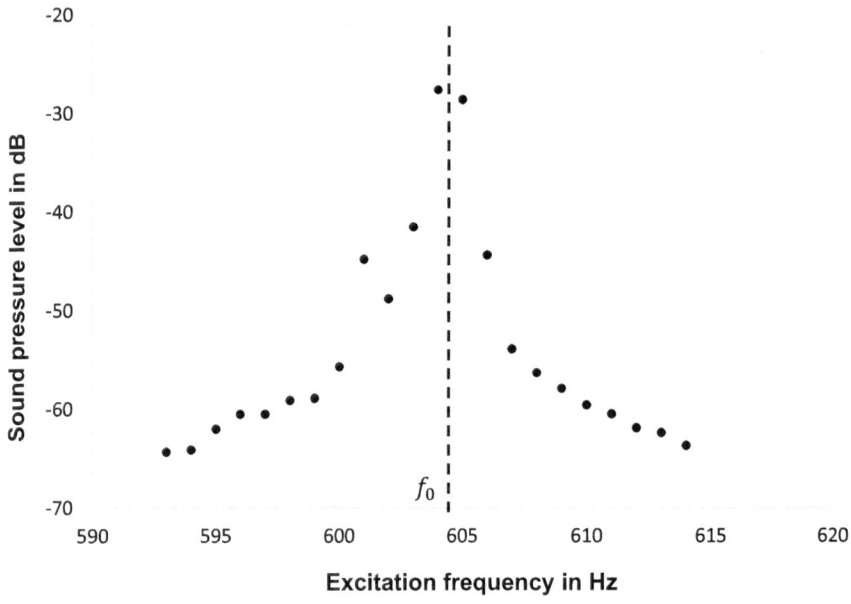

Fig. 3.12 Graphical representation of the series of measurements (sound pressure level not calibrated)

Natural frequency, forced vibrations and resonance

Once a body has been excited to vibrate, it performs free vibrations (natural vibrations) at a certain frequency. This frequency depends solely on the body itself and is therefore called the *natural frequency*. For example, the natural frequency of a pendulum depends on its length or that of a spring pendulum on the mass of the vibrating body. As a result of friction losses, the amplitude of a free vibration constantly decreases and the body takes its rest position after some time. If, on the other hand, the vibration of a body is maintained by periodic energy supply from the outside, it performs so-called *forced vibrations*. You know this, for example, from the children's swing, where you regularly supply energy to the vibrating system by pushing or by skillfully shifting your own center of gravity. The frequency of the energy supply is called the *excitation frequency*.

The amplitude of an excited vibration depends on the excitation frequency and reaches its maximum in the case that the excitation and natural frequency correspond to each other—this is then referred to as *resonance*.

Under certain circumstances, the excited vibration can even lead to the destruction of the body. A prominent example of such a resonance catastrophe is the collapse of the Tacoma Narrows suspension bridge in the US state of Washington. The torsional vibration excited by the wind led to the collapse of the bridge on 11/07/1940, just four months after it was put into operation. To prevent bridges from being excited to forced vibrations by marching in step, this is strictly prohibited by the German Road Traffic Regulations (StVO). It states in paragraph (6): "Marching in step on bridges is not allowed."

Sound Speed for Beginners and Advanced

Experiment 18: Measurement at the "Palatinate Tube"

Perhaps you still remember from your school days a classic method to determine the speed of sound. One person stands on the sports field with a starter's gun ready. At a precisely known distance—the 100-m track is suitable for this—other people stand with stopwatches. As soon as they see the starter's gun being fired, they start the measurement. When the "bang" arrives, the measurement is stopped and the speed of sound is calculated from the time difference and the distance between the starter's gun and the stopwatch. (The speed of light of the optical signal can be safely neglected here.) This is a nice measurement method because the phenomenon of time delay can be impressively experienced here. However, this measurement is also characterized by larger uncertainties. In particular, the human reaction time makes the result inaccurate.

Thanks to technical developments and interesting smartphone apps, the speed of sound can now be measured in a variety of ways, very simply and above all more accurately. We start with a "beginner's method", the theoretical foundations of which were already pointed out when uncorking (see Experiment 1).

For the measurement, we only need one or two smartphones and a tall cylindrical glass (Vogt et al., 2014). Such glasses are an essential part of everyday equipment in the Palatinate (Germany) as wine glasses. Therefore, we have given them the name "Palatinate Tube".

The measurement of the speed of sound is basically done in the same way as when uncorking (see Experiment 1). There, a "wide range" of acoustic frequencies was offered to the resonance chamber in the bottle neck through friction and air flow when pulling the cork. This is not possible with a drinking glass. But we can use a smartphone with an app installed to generate "white noise" (see info box). We now "rumble" into the glass with it (Fig. 3.13). Of the frequencies contained in the white noise, the one that is characteristic for the glass geometry (especially the height) and the gas contained in it (usually air) is amplified as a resonance frequency. We will not be able to hear the amplified frequency without aids. But again, a smartphone app can help. This must be able to display a frequency spectrum. Often this is referred to as an "FFT spectrum" (see info box). Ideally, an app can be used that can do both at the same time. We used the "Audio Kit" app for this. Alternatively, the measurement can also be carried out with two smartphones, one of which generates the noise and the other records the

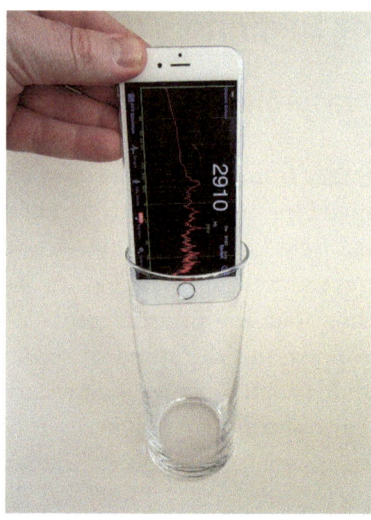

Fig. 3.13 Determination of the speed of sound with the "Palatinate Tube"

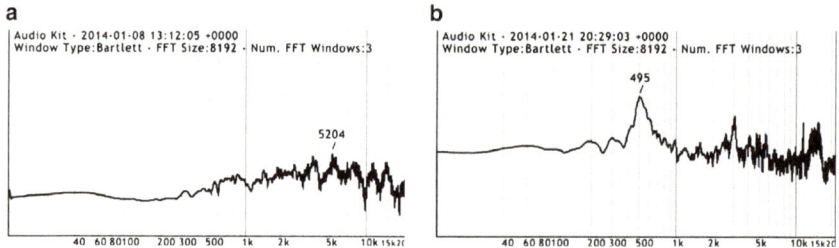

Fig. 3.14 Screenshots for the measurement of the speed of sound; white noise without glass (**a**), resonance frequency in the "Palatinate Tube" (**b**) (App: Audio Kit)

frequency spectrum, or the white noise is generated by crumpling a piece of paper. The results of our measurement are shown in Fig. 3.14. Screenshot (**a**) shows the frequency distribution of the white noise without glass. It "fidgets" a bit, but no clearly pronounced individual frequencies can be seen. Screenshot (**b**) then shows the measurement of the frequencies while the smartphone is rumbling into the glass. Here a clearly highlighted frequency at 495 Hertz is recognizable.

With this measurement result and the basics from Experiment 1, we can now determine the speed of sound. The glass used for the measurement has a height L of 15 cm and a radius R of 3.7 cm. This results in the speed of sound in air as follows:

$$c_{\text{air}} = 4f_0(L + \Delta L) = 4 \cdot 495 \text{ s}^{-1} \cdot (0.15 \text{ m} + 0.61 \cdot 0.037 \text{ m})$$

$$c_{air} \approx 342 \; \frac{m}{s}$$

The measurement was carried out at a room temperature of 24 °C, which theoretically would expect a speed of sound of 346 m/s. Therefore, this simple measurement method achieves a surprisingly good result.

"White Noise"

This is how the superposition of many frequencies is described in acoustics. The power density of the frequencies is constant. This means that there is no highlighted value of a sound pressure level for any frequency. In a frequency spectrum, an ideal white noise would therefore represent a constant (in the diagram as a horizontal line).

An analogy to acoustic white noise in optics is white light. This light also consists of the superposition of many optical wavelengths. However, the constancy of power density does not apply to light that we perceive as "white".

FFT Sound Analyses

FFT here stands for *Fast Fourier Transform* or *quick Fourier transformation*. This is a mathematical "trick" in which a given function (here the superposition of all acoustic frequencies of a sound) is developed according to a system of basic functions. With a sufficiently fast computer, it is thus possible to determine the amplitudes of small frequency intervals of any acoustic signals and to represent them as a spectrum, e.g. in Fig. 3.14.

FFT analyses are also used in everyday life in a variety of ways: from the acoustic "fingerprint" in forensics to the evaluation of the singing voice in karaoke programs.

Experiment 19: The Wine Glass as a Helmholtz Resonator

Of course, it could well be that you do not have a spritzer glass in front of you, but perhaps you prefer a red wine. Even in this case, you do not have to forego a determination of the speed of sound. The necessary calculation is only slightly more complex than with the "Palatinate tube" and goes back to a universal scholar of the 19th century, *Hermann von Helmholtz* (1821–1894).

The procedure of the measurement basically follows Experiment 18 in the previous section. We now need a (empty) bulbous glass and a smartphone (Monteiro et al., 2015). If this device is equipped with an app that can simultaneously generate a white noise and perform a frequency measurement, no further smartphone is required.

While the smartphone generates a white noise, it is slightly dipped into the glass with the frequency measurement activated and the side where the microphones are located. From the mixture of many frequencies in the white noise, the glass selects its resonance frequency and amplifies it. Analogous to the previous experiment, this frequency is again displayed as a clear peak by the smartphone (Fig. 3.15) and with its help, the speed of sound can be inferred. However, the wine glass is now not considered as a simple one-sided open tube, but as a so-called *Helmholtz Resonator* (see info box). For such a cavity resonator without a neck, the resonance frequency f_0 results from the following equation (Trendelenburg, 1950, p. 225):

$$f_0 = \frac{c}{2\pi} \cdot \sqrt{\frac{2R}{V}}$$

In addition to the resonance frequency measured by the smartphone, we therefore need the opening radius R of the wine glass and the volume V of the cavity. While the radius can be quickly measured with a ruler, the volume can best be determined with a measuring cup. By the way, one quickly marvels at this. A whole bottle can certainly fit into some red wine glasses (which would probably not be well received in social practice …).

Since we want to determine the speed of sound c here and have measured the resonance frequency, the equation is rearranged and the values determined on the glass from the example in Fig. 3.15 are inserted:

Fig. 3.15 Determination of the speed of sound with bulbous glasses

$$c = 2\pi f_0 \cdot \sqrt{\frac{V}{2R}} = 2\pi \cdot 596 \text{ s}^{-1} \cdot \sqrt{\frac{0.7 \cdot 10^{-3} \text{ m}^3}{2 \cdot 0.04 \text{ m}}} = 350 \text{ ms}^{-1}$$

From the frequency measurement, a speed of sound of approximately 350 m/s should therefore result. The experiment was carried out at a room temperature of 25 °C and thus a speed of sound of approximately 346 m/s is expected. For such a—from a physical point of view—rough experiment, the result with a relative error of about 1% is surprisingly good!

What is a Helmholtz resonator?

In 1859, the physiologist and physicist *Hermann von Helmholtz* (1821–1894) developed a hitherto missing theory of cavity resonators that could explain acoustic relationships between the geometry of a gas-filled cavity and its resonance frequency. If such a cavity with as simple a geometric shape as possible (Fig. 3.16a) is acoustically excited by different frequencies, then the frequencies predicted by the Helmholtz theory show an amplification.

In historical practice, the resonators were used, among other things, for frequency analysis. *Helmholtz* himself described it in such a way that he covered the narrow opening with a still liquid, but no longer painfully hot wax and then inserted it into the ear canal. The cooling wax then closed the remaining gaps between the ear and the resonator. The other opening was now exposed to noises of different frequencies. The resonance frequency predetermined by the dimensions of the resonator was noticeably amplified.

The cavity resonators shown in Fig. 3.16b, c and d are less used in everyday practice in acoustic contexts. However, it can be shown that they, with more or less restrictions, certainly satisfy the Helmholtz theory (see also Experiment 20).

Experiment 20: Wine Bottles as Helmholtz Resonators

After demonstrating that bulbous wine glasses serve very well as acoustic cavity resonators according to Helmholtz's theory, we now turn to the emptied bottle. Can we also apply this theory to wine bottles? If so, we should be able to predict the typical frequency audible when blowing into an empty wine bottle. Let's check this out!

A common bottle shape is the Bordeaux bottle, as shown in Fig. 3.16c. Its characteristics are the so-called shoulders and—of particular interest here—the long bottle neck with a constant diameter. For the prediction of the frequency, the bottle should be considered as a Helmholtz resonator. For the

a   b   c   d

Fig. 3.16 Historical (**a**) and unconventional Helmholtz resonators (**b–d**)

fundamental frequency of such resonators with a long neck (compared to the opening), the following equation applies (Trendelenburg, 1950, p. 225):

$$f_0 = \frac{c}{2\pi}\sqrt{\frac{\pi \cdot R^2}{V \cdot L}}$$

Here, c is the speed of sound, for which here (at an ambient temperature of 22 °C) 345 m/s is used.

R is the radius of the bottle opening and is 1 cm. L as the length of the bottle neck is 8 cm here. We of course also know the volume V of the bottle: 0.75 l. Measured precisely, the oscillating air volume is however 0.79 l.

If you insert all quantities into the equation for the frequency, you get the theoretically expected fundamental frequency of $f_0 = 122$ Hz. For the experimental verification of this prediction, we again need a smartphone or another device with a frequency measurement capability. In this case, an app is used that outputs the measured fundamental frequency as well as the musical tone (e.g., *Spaichinger sound analyzer*). Now we let the bottle sound. The measurement for the blown Bordeaux bottle is shown in Fig. 3.17. There, the peak frequency—which is our sought-after fundamental frequency—is given as approximately 115 Hz. Thus, our theoretical prediction is not so bad. The relative error is on the order of less than 6%.

Friends of Franconian wines have other acoustic resonators available in the form of the "Bocksbeutel" (Fig. 3.16d) after enjoying them. Bocksbeutels are noticeable for their bulbous and flat bottle shape and are indeed almost a privilege of the Franconians. In their mountainous terrain, the bottles have the invaluable advantage during a picnic in the open air that

Fig. 3.17 Fundamental frequency of a blown wine bottle (App: Spaichinger sound analyzer); top: measured value, bottom: music note matching the measured frequency

they do not unintentionally move downhill. Of course, the uncomplicated acoustic experiment can also be carried out with Bocksbeutels. With the same volume (0.75 l), a neck length of 7 cm is measured on our sample bottle. The inner radius of the neck is 1 cm, as with the Bordeaux bottle. According to the equation for the frequency used above, which also applies here, a fundamental frequency of 135 Hz is to be expected when blowing into the empty bottle. The measurement on our sample bottle here gives a frequency of 124 Hz. This value also agrees reasonably well with the prediction, with a relative error of 8%.

Both experiments thus show us that wine bottles can be roughly considered as Helmholtz resonators, even if they do not follow Helmholtz's theory in an exact manner.

There's Music in It—Glasses and Bottles as Instruments

Experiment 21: Wine Glass Harmonica

Pink Floyd used them several times in live concerts (e.g., in the title *Shine On You Crazy Diamond*), and they are also sometimes heard on stage in classical music—played virtuosically: singing wine glasses (Fig. 3.18). And if you and your guests are currently taking a drinking break, then gather all the glasses, have a bottle of water and a tuning device ready, and then you can start. Of course, there is a lot of physics behind this not-so-everyday instrument, which we will explore in the following.

Fig. 3.18 "Singing wine glasses" as a musical instrument

One method to make glasses ring is to glide over their edge with a moistened finger. With just a little practice, one quickly finds the right pressure and a—in the truest sense of the word—crystal clear and delicate sound can be heard. How does this happen? A hint can be obtained by observing the water surface in a glass filled with some water while rubbing the edge of the glass. It shows fine ripple-like wave patterns, especially at the edge of the glass (Fig. 3.19a). Obviously, the glass is stimulated to vibrate quickly by the process and transmits these vibrations to the water. The glass is excited to vibrate by the sliding finger, which actually does not glide so "smoothly", but alternately sticks and slides again. We also know this from the terribly squeaky chalk on the school blackboard. In the vibration theory of physics, this is referred to as the *stick-slip effect* (see info box). The water on the moistened finger has the task of reducing the friction to the right measure. A closer look at the wave patterns of the water in the rubbed glass shows us that these wave ripples move in the same direction of rotation as the finger.

Another method to make the glass ring is to strike it with a wooden stick or a small spoon. If you do this at a table with many guests, all eyes will turn to you in anticipation of a speech, despite the noise level of the party. This shows how penetrating the sound can be. Here too, wave patterns become visible again on the water surface in the glass filled with some water after

Fig. 3.19 Vibrations made visible; rubbed glass edge (**a**), struck glass (**b**)

striking it. However, in this case they appear as concentric wave crests and troughs (Fig. 3.19b).

If one judges by ear the rubbing of the finger and the striking—applied to the same glass, one will find that the sounds are similar, but not exactly the same. Try it out! Looking at the frequency spectra, it can be seen that the fundamental frequencies are the same or almost the same in each case. The glass vibrates in both cases at its resonance frequency, which is built in by its geometry and glass thickness, but which can be changed by the amount of liquid contained in the glass. However, the spectra of the overtones differ in the two types of excitation the overtones (Fig. 3.20). The clear dominance of the fundamental tone and the comparatively few overtones produce the very special sound when rubbing the glass.

To illustrate the creation of the fundamental tone of a wine glass, we can refer here to a simple "coffee mug model" (Vogt et al., 2015; Fig. 3.21). The vibrating glass wall forms places of maximum and minimum movement, which are referred to as "vibration bellies" and "vibration nodes". In the simplest mode of vibration, which corresponds to the frequency of the fundamental tone, these are four opposing and mutually vibrating surfaces, separated by four node lines. Since we see further frequencies in the spectrum, which correspond to the overtones, there must also be further patterns of bellies and node lines. These are superimposed on the fundamental vibration. In this respect, wine glasses and bells are very similar (see also Experiment 15).

The differences in the wave patterns on the water surfaces in Fig. 3.19 can be explained by the fact that the patterns forming on the glass surface from vibration nodes and nodal lines migrate with the circling finger when

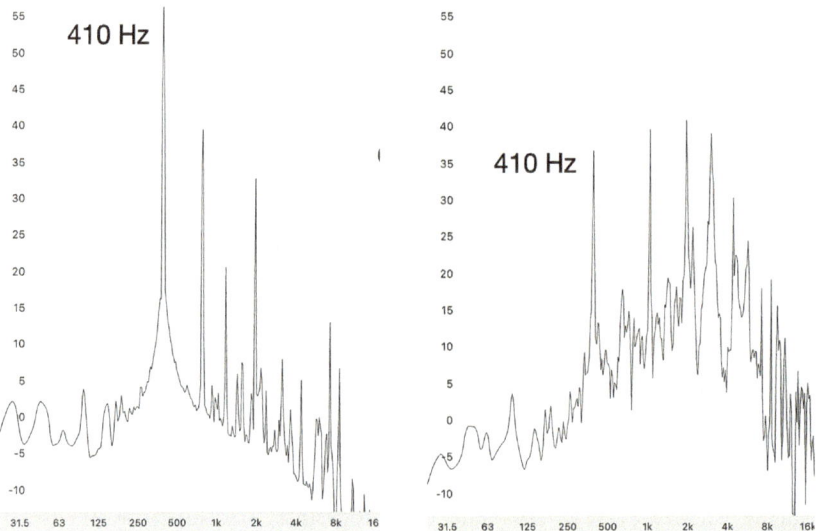

Fig. 3.20 Comparison of the different sounds produced on the same glass with unchanged water filling; "rubbed" wine glass (**a**), struck wine glass (**b**)

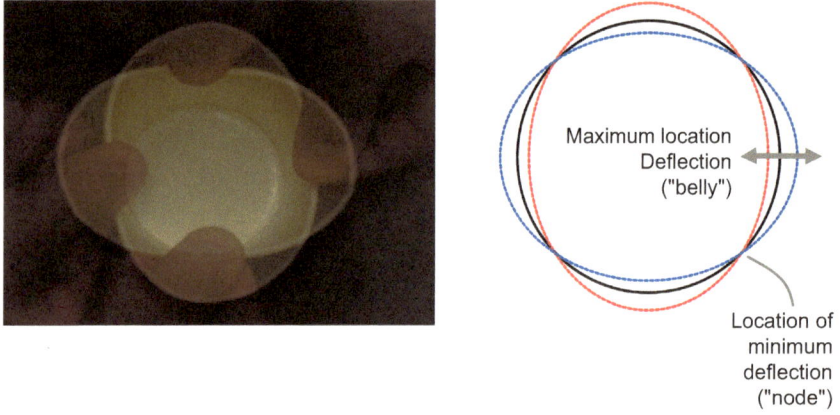

Fig. 3.21 "Coffee mug model" for the fundamental vibration of a wine glass

rubbing. However, if the glass is struck, the patterns on the glass surface remain static.

The starting point of the previous considerations was to assemble several wine glasses into one instrument. For this, we now need to address the pitch of the vibrating glasses. Let's assume that only one type of wine glass is on the guest table. Then the tuning of the glasses must be done through different filling with a liquid. We have measured a large Burgundy glass as an example, first without any filling and then with a maximum filling, where a sound was still clearly audible. This glass therefore has its upper frequency limit at 398 Hz in the empty state and its lower limit at about 260 Hz. The glass was not completely filled, but the sound quality drastically deteriorates with further filling. With five glasses of this type, the sequence of musical notes C4 (262 Hz), D4 (294 Hz), E4 (330 Hz), F4 (349 Hz) and G4 (392 Hz) can be achieved by appropriate filling. At least: The folk song "Little Hans" is already possible and the bottle pan flute proposed in the following experiment would harmonize perfectly with this wine glass harmonica!

Finally, the question remains to be clarified why the pitch of the rubbed or struck glass decreases with the filling level of a liquid. For this, we return to the walls of the glass stimulated to vibrate. If the glass is empty, it can vibrate at its resonance frequency. With a filling in the glass, the vibrating glass walls have to work against the inertial resistance of the liquid. The vibration thus becomes slower and the frequency decreases. The more liquid the glass contains, the lower the tone of the fundamental frequency becomes.

By the way, a historical note: Already around the year 1760, the statesman and natural scientist *Benjamin Franklin* (1706–1790) invented a glass harmonica that worked exactly according to the friction principle of the wine glass harmonica described here. Mounted on a common axis were several dozen glass bells, stacked in such a way that their edges could each be played (rubbed) with the finger. *Franklin's* invention on this instrument consisted of the foot drive, with which all the glass bells could be set in motion together (Fig. 3.22).

Stick-Slip Effect

The Stick-Slip Effect describes the interplay of static and Sliding friction between the surfaces of two solid bodies rubbing against each other. For this "jerky friction" effect to occur, the static friction force F_{sf} must be significantly greater than the Sliding friction force F_{kf}. This results in the following for the corresponding friction coefficients:

$$\mu_{sf} \gg \mu_{kf}$$

In the system of the rubbing bodies, the alternating sticking and sliding leads to the excitation of vibrations, which are often radiated as sound from the surfaces. The fact that apparently—as seen in the example of the rubbed wine glass—precisely the natural frequency of the glass is hit, should be very surprising. In fact, a mixture of different frequencies is created by the Stick-Slip Effect.

Fig. 3.22 Glass harmonica with foot drive (CC-BY-SA 4.0, Historisches Museum Frankfurt (X25198), photo: Uwe Dettmar)

> However, the physical system "wine glass" amplifies its natural frequency as a resonator, resulting in the clearly audible resonance.
>
> The Stick-Slip Effect is desired when bowing the string of a musical instrument. It is undesirable, for example, when door hinges squeak. In such cases, the two rubbing solid surfaces are separated by adding a lubricant.

Experiment 22: Wine Bottle Pan Flute

Over time, wine lovers accumulate a number of empty bottles. As we have already seen in Experiment 20 ("Helmholtz Resonators"), these can also be suitable for experimenting. This is exactly what we want to build on now. But this time, several empty bottles should be used. With these and the findings of the Helmholtz Resonator experiment, we can make excellent music, and together with the previously discussed glass harmonica, we could almost speak of an orchestra.

We are talking here about a "wine bottle pan flute". In the classification of musical instruments, this would be assigned to the so-called *vessel flutes*, a type of instrument used very early in cultural history, namely already in the Neolithic period.

To limit the effort and simplify the later handling, we present here a 5-tone bottle flute. The basis of the instrument is—this is also a simplification of later calculations—the fivefold same model of a standard wine bottle in Bordeaux shape (the one with the "shoulders"), for which the fundamental frequency has already been calculated (121 Hz) and measured (115 Hz) in the completely emptied state (see Experiment 20). This would almost correspond to the musical note A2 sharp or A#2 and would be the lowest tone that can be achieved by blowing into this bottle.

So, how large is the range of tones of this bottle? We could calculate the highest achievable tone again with the previously given equation. For this, the volume of the oscillating air in the bottle would have to be minimized. But there is also a simpler way. Let's assume that the bottle is filled to the extent that only the 8 cm long bottle neck is available as a resonator. For this case, we already know the fundamental frequency from the first corkscrew experiment, which is 925 Hz (Table 1.1). The highest usable tone for a bottle pan flute should therefore have a frequency of less than 900 Hz and the range of tones for a 5-tone instrument is thus easily sufficient. If you want to move in the lower range of the tone range, the following tones (with associated fundamental frequencies) are available: C3 (130.8 Hz); D3 (148.8 Hz); E3 (164.8 Hz); F3 (174.6 Hz) and G3 (196.0 Hz).

Now the bottle flute needs to be tuned. Those with a musical ear can be successful by successively filling, blowing, and listening. The good ear could also be replaced (or supported) by a suitable smartphone app. Nevertheless, the method of calculation should not be neglected here. We rearrange the equation of the bottle-Helmholtz resonator for the desired air volume:

$$V = \frac{c^2 \cdot R^2}{4\pi \cdot f_0^2 \cdot L}$$

Where f_0 is the required fundamental frequency for the respective desired tone.

For the note c with a rounded 131 Hz, this results in an air volume of $V = 686$ cm³. To reduce the 790 cm³ of air in the empty bottle, 104 cm³ of water would therefore have to be poured into the empty bottle.

Here too, the control measurement of the generated tone shows deviations from the calculations with a relative error of about 6%. However, the calculated fill volumes can be used well as rough guidelines. Table 3.2 provides the example calculations compared to the measurements for the bottle used here.

Table 3.2 Calculated and measured tuning of a bottle pan flute for a temperature of 21 °C

Tone	Fundamental frequency (rounded) in Hz	Calculated fill volume in ml	Measured fundamental frequency in Hz	Relative error in %
C3	131	104	126	3.8
D3	147	245	139	5.4
E3	165	358	161	2.4
F3	175	406	168	4.0
G3	196	484	194	1.0

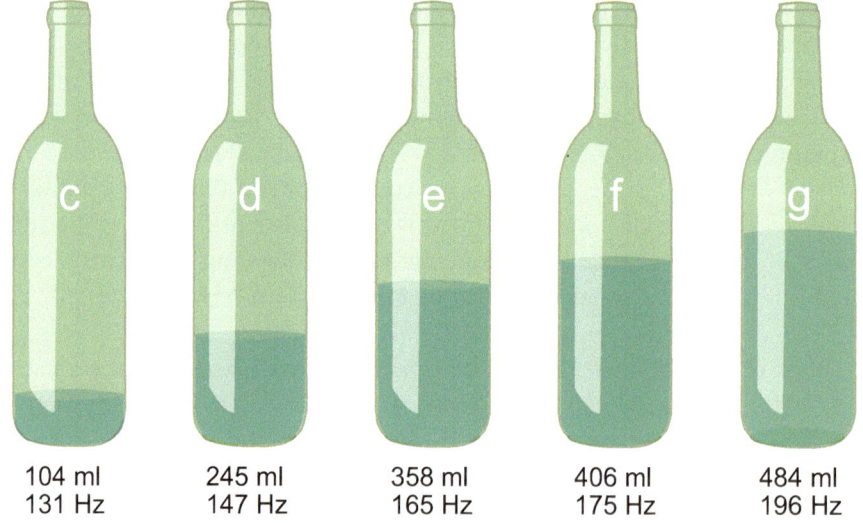

| 104 ml | 245 ml | 358 ml | 406 ml | 484 ml |
| 131 Hz | 147 Hz | 165 Hz | 175 Hz | 196 Hz |

Fig. 3.23 Fill volumes, tones and fundamental frequencies of blown bottles using the example of a Bordeaux wine bottle

The values given in Table 3.2 and Fig. 3.23 were calculated or measured at a temperature of 21 °C. This note shows that our bottle pan flute tuned in this way is temperature sensitive. In the calculation equation, this dependency is hidden in the speed of sound c, which in turn depends on the temperature (see info box).

Speed of sound and temperature

The propagation of acoustic waves always requires a medium. This can be a solid, a liquid or a gas. For experiments with blown bottles, the vibrating air in the bottle plays the decisive role. For gases like air, it is true that they become harder to compress as the temperature rises. For the propagation of sound in gases, compressibility is a crucial factor. The harder a gas is to compress, the faster the sound vibrations spread. The following proportionality applies:

$$c \sim \sqrt{T}$$

Where T is the absolute temperature (measured in Kelvin). This results in sound propagating faster in warmer air. For a simple estimate of the speed of sound in air at different temperatures, the following relationship can be used (Lüders & von Oppen, 2008, p. 525):

$$c = (331.3 + 0.6 \cdot \vartheta/°C)\frac{m}{s}$$

Where $\vartheta/°C$ is the numerical value of the Celsius temperature. For air at 21 °C, this results in a speed of sound of about 344 m/s.

4

Enjoy with All Senses: Optical Phenomena with Wine Glasses

If the previous chapter was titled "The Ear Drinks Too" in its headline, the same undoubtedly applies to the eye. Not without reason does the saying go that a person who has intensely engaged with wine has looked too deeply "into the glass". But that's exactly what we intend to do in this chapter. Once again, glasses of various shapes and different glass fillings come into play. What does the world look like through a bulbous glass? Does red wine actually make the world red? And if there's too much red, then we simply make a Blanc de Noirs out of it. Or do you fancy a cigar with your wine, but don't have a lighter at hand? No problem, we'll find a solution …

Wine Glasses as Lenses

A glorious summer afternoon in a medieval town. You sit with a rosé and realize before the first sip that the world is contained in the glass (Fig. 4.1). But wait, it's upside down! How does that happen? With some interesting observations and experiments, we delve into the optical properties of filled wine glasses.

Experiment 23: Burgundy Glass—Spherical Lens and Cobbler's Ball

As the name of the cobbler's ball suggests, it once served shoemakers and other craftsmen for economical, but precise lighting. For this purpose,

L. Kasper and P. Vogt, *Uncorking the Physics of Wine*, https://doi.org/10.1007/978-3-662-68759-8_4

Fig. 4.1 The world in the glass is upside down

a ball filled with water and a candle were positioned in such a way that it produced a very bright spot of light at the focal point of the ball, which is just outside the ball. The shoemaker could thus see his seams much better (Fig. 4.2).

To experience the effect of a cobbler's ball, we can also resort to a spherically curved wine glass. Candles are often part of a beautiful ambiance anyway, so the experiment is almost set up. However, it should be a white wine or—scientifically a bit more correct—a wine glass filled with water. Physically speaking, bulbous glass shapes then approximate spherical lenses, which are described in more detail in the following info box.

If you have really filled a "spherical" wine glass with water, you can literally turn it into a burning glass in the sunlight, with which you can actually manage to ignite a piece of paper held behind the glass.

Fig. 4.2   A cobbler's ball as a historical craftsman's lamp

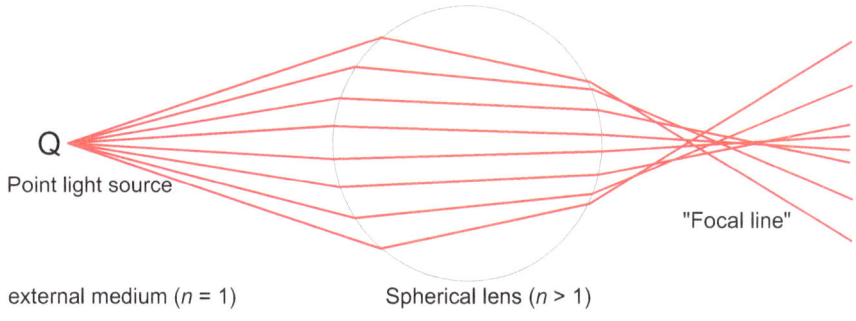

Fig. 4.3   Double refraction of divergent light through a spherical lens

Physics of Spherical Lenses

The term spherical lenses refers to light-permeable spherical bodies that differ from the medium surrounding them by their optical refractive index. In other words: Light propagates within such a spherical lens at a different speed than outside. When light hits the spherical surface of the lens from the outside, it is refracted. When it exits the lens again, another optical refraction occurs. Fig. 4.3 shows a solid glass spherical lens (refractive index approx. 1.5), which is hit by light from a point source from the air (refractive index approx. 1.0).

In the model construction in Fig. 4.3, something like a focal point can be seen behind the lens. Indeed, paper can be easily ignited with a glass ball and sunlight. However, the focal point is not a real point. Spherical lenses have significant lens errors, especially the *spherical aberration,* which turns the focal point into a line.

Where is the focal point of a spherical lens? The answer depends on three parameters. These are the refractive indices of the lens material and the surrounding medium, and the radius of the spherical lens.

$$f_{lens} = \frac{n_{lens}}{n_{lens} - n_{medium}} \cdot \frac{R}{2}$$

This focal length is calculated from the center of the spherical lens (principal plane of the lens). The equation only applies to light that falls on the lens near the central axis due to the aforementioned spherical aberration.

If the spherical lens is in air, then $n_{medium} = 1$. Using the above equation, it can be seen that the focal point is outside the sphere as long as the refractive index of the lens material remains less than 2 (which is the case with a glass ball and a glass filled with water).

The course of light within a spherical lens can be made visible in the following way: Instead of a glass ball, a wine glass filled with water (refractive index approx. 1.3) is used. The water in the experiment in Fig. 4.4 was clouded with a few drops of milk for better visibility of the light passage. The light

Fig. 4.4 Refraction when parallel light enters a bulbous wine glass

source is an old slide projector, which produces a nearly parallel and very bright beam of light.

Spherical lenses have historically been used primarily as lighting aids. However, they are still used today in miniature optical systems as objective lenses, e.g., for optical sensors or for coupling light into fibers.

Experiment 24: Fun with Cylinder Glasses

Not all wine glasses are bulbous and we have already used cylindrical glasses for acoustic purposes—e.g. the "Palatinate Tube" as a spritzer glass. If such glasses have a roughly good cylindrical shape and the glass wall is not too thick, interesting observations can also be made from an optical point of view. However, the glass must be filled for this. Remember: Optical phenomena come before wine enjoyment, after emptying the glasses it is better to experiment acoustically with them!

In the following, the properties of suitable glasses as cylindrical lenses are to be examined in particular. We call lenses that have a cylindrically shaped surface cylindrical lenses. Unlike in spherical lenses, light refraction in cylindrical lenses occurs in only one dimension.

Many people who wear glasses or contact lenses may not be aware that the usually spherically ground lenses are often overlaid with a more or less pronounced cylindrical cut. This is always the case when the eye, in addition to being far- or shortsighted, also has a so-called *astigmatism* (a rod vision).

While optical components in almost all cases only have the shape of a cylinder segment, the "most cylindrical" lens you can think of is an optical lens whose shape is a complete cylinder. We find such a thing in the household in the form of cylindrical transparent glasses or bottles. And so we are back at the "Palatinate Tube". However, such a glass only becomes a cylindrical lens through its filling. Of course, besides spritzer, water can also be used, whose refractive index we know well ($n = 1.3$).

In the experiment shown in Fig. 4.5, a vertical inscription is located directly behind a glass cylinder filled with water.

The inscription in Fig. 4.5a is easy to read, the letters appear somewhat widened. So there is a magnifying effect—but only one-dimensional, and only in one direction perpendicular to the cylinder axis.

If we slowly pull the cylinder away from the inscription, i.e. towards us, after a short "overthrow" the inscription no longer appears enlarged, but

Fig. 4.5 View through a cylindrical glass; glass directly in front of the inscription (**a**), glass somewhat removed from the inscription (**b**)

again clearly readable (Fig. 4.5b). Surprisingly, in our experiment, the green word seems to have "survived" the overthrow process unscathed, while the red components are readable in mirror image! It is certainly not an effect of color and on closer inspection it becomes clear that all green letters have a vertical axis of mirror symmetry. They have also swapped their sides, but we don't really notice.

Try to perform some optical tricks with such glasses yourself. Move— while looking through the glass—an object from one side into the field of view behind the glass. Observe where the object first appears.

Cylindrical lenses

Unlike spherical lenses, cylindrical lenses focus only in one direction, i.e., they do not bundle light in a focal point, but in a focal line parallel to the cylinder axis (Fig. 4.6).

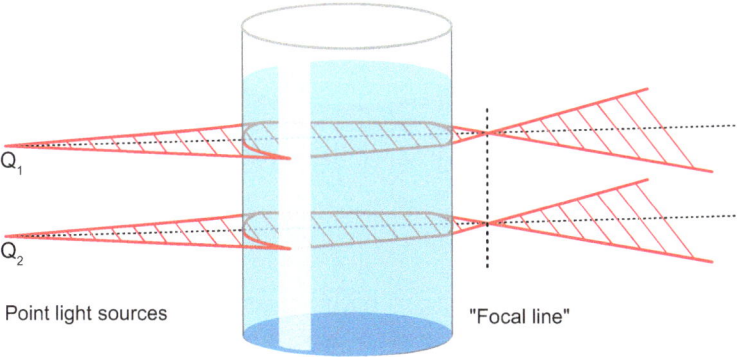

Fig. 4.6 Selected optical ray paths on a cylindrical lens

For the position of the focal line behind the cylindrical lens, in the case of lenses with a circular cross-section, e.g. glasses filled with water, the equation for spherical lenses applies again (see Experiment 23).

With two crossed (perpendicularly arranged) cylindrical lenses each with the same radius of curvature, images can be achieved that correspond to those of a spherical lens. That is, they then produce not a line image for an object point, but again an image point.

Looking Through Glass

Experiment 25: Red Wine as a Color Filter

So far, our views through variously shaped wine glasses have focused on their geometric shape and the associated refractive properties they have when filled with a clear liquid. In the following, the shape of the glasses should play no role. We now turn more to the profound red color of the wine. As is well known, the eye also drinks. Thus, the perceived color often influences our judgment about a wine even before the first sip. We tend to attribute higher quality to particularly dark red wines and those with a red-violet color shade. At any rate, it should preferably not have any brownish accents. Overall, the topic of "color" is highly complex. Here, in addition to the actual and weather-related fluctuating content substances, the methods of wine preparation, the later maturation processes, the degree of maturity, the storage of the wine with varying oxygen contact, and the temperatures interact in a variety of ways.

To find a clear approach, we start with a simple question that can be easily decided experimentally:

What happens when you shine through red wine with two differently colored light sources, such as a red and a green laser pointer (Fig. 4.7)? Do both light beams penetrate the wine? Before we investigate this experimentally (Kasper & Vogt, 2022), we also ask why red wine appears red to us at all. Or, more generally: What actually makes up the color impression of an object that we look at?

We assume that the wine is not a self-luminous object like a star or a lamp (if it were self-luminous, you should be very careful with the tasting). We can see objects that do not shine themselves because they are illuminated by light sources. A glass of wine—viewed in daylight—is illuminated by the sun. Sunlight, as white light, contains the entire color spectrum that we can perceive. Expressed in wavelengths, this is approximately the wavelengths from 400 nm (blue-violet), which border on the invisible ultraviolet, to 780 nm (red), where the transition to the also invisible infrared is.

The red wine absorbs this white light mixture and absorbs part of this radiation. It is due to its very specific properties, such as the content of red pigments, mainly from the skin of the berries, that the wine only radiates very specific light colors (certain wavelengths of light). The color impression finally results from the mixture of the radiated colors.

Illuminating a layer of red wine with differently colored light now leads to the concept of a color filterand thus to the so-called *subtractive color mixing*. This term already suggests it: A color filter "subtracts" certain wavelength ranges of the incident light and only lets a part through.

In the case of red wine, it is a wavelength range that lies in the visible red. The *polyphenols* in red wine, and especially the *anthocyanins*, largely determine its color. Depending on the pH value, anthocyanins can be energetically excited by visible light. This absorbs green light in the wavelength range around 520 nm. This results in our visual impression for this pigment lying precisely in the complementary color red, and thus the entire liquid appears red to us.

The result for the laser pointer experiment is thus clear. The green light of the laser (in our case with a wavelength of 530 nm) is absorbed by the red color pigments. Behind the cuvette filled with red wine in the experiment (Fig. 4.7), no green light point is visible on the screen. The red laser light, on the other hand, penetrates the red wine almost undisturbed and causes a red light point on the screen. Red wine in a suitable transparent container is therefore well suited as an optical filter.

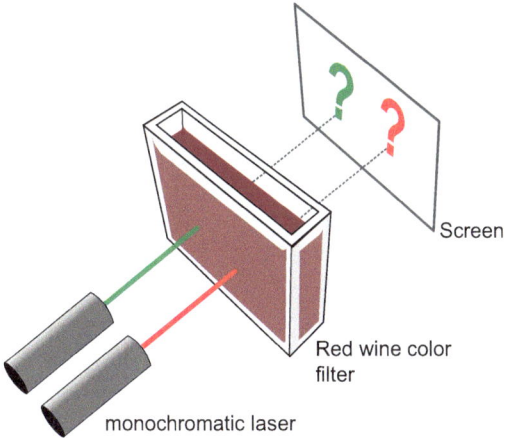

Fig. 4.7 Laser pointer with green and red light hitting red wine

---

Color Filters in Optics

Transmittance
Optical filters refer to transparent solid substances (e.g., glass) or liquids that absorb or reflect at least part of the spectrum. The ratio formed by incident and transmitted light is referred to as the *transmittance* τ. In the case of monochromatic light, this is the *spectral transmittance* $\tau(\lambda)$. It must always hold that: $\tau \le 1$

Absorption Filters
Optical filters that are based on absorption are called *absorption filters*. With these filters, almost all light components that are not transmitted are absorbed by the filter material. These materials contain colored pigments that are added to the glass melt or the liquid. In the case of red wine, these are primarily the anthocyanins, which absorb the green component of the light.

In addition to absorption filters, *interference filters* are also used in optics, but they will not be discussed here.

The color filter experiment can naturally be extended to other colors. For this, it is best to look through a pair of "red wine glasses" or a well-filled glass of red wine! All shades of red are transmitted, while colors appear darker, even black, the less red they contain. In Fig. 4.8, the cuvette filled with Merlot from the previous experiment was simply held in front of a camera lens. You can see how even the slight green component of the green-yellow lime in the middle, when viewed through the red wine color filter, creates an almost black color impression, while the yellow of the

Fig. 4.8 Various colored citrus fruits seen through a "red wine color filter"

ripe lemon and the orange of the mandarin obviously show a high red component.

Experiment 26: A Look into the Glass with the Infrared Camera

It seems quite simple: white grapes are used to make white wine, red grapes to make red wine. But how do you make a rosé or even a Blanc de Noirs (French for "White from Blacks")? White grapes inevitably yield white grape juice when pressed, which is processed into white wine through fermentation and appropriate aging—there's no disputing that. The situation is different with red grapes, whose pulp and juice are often also white. The colorants, the anthocyanins already mentioned in Experiment 25, are initially mainly located in the berry skin (Fig. 4.9).

Depending on the processing of the mash, these pigments more or less strongly enter the wine. If you want to produce a red wine, you let the juice

Fig. 4.9 Even with most red grape varieties, the juice and the pulp of the berries are white, the colorants are mainly in the skin

ferment on the berry skins, whereby the resulting alcohol dissolves the dyes from the skin (so-called mash fermentation). The pressing of the mash and thus the separation of juice and skin takes place after the fermentation. To produce a rosé, the contact time of juice and berry skins is limited to a few hours, so that less color transfers into the juice. For the production of a Blanc de Noirs, juice and berries should be separated as quickly as possible after the harvest, i.e., the grapes are pressed immediately. How well the white wine production from red grapes actually works depends on the coloring of the juice and thus on the grape variety. The pulp of Regent grapes, for example, already has a reddish tone, so this grape variety would be unsuitable for the production of a white wine. This is quite different with Pinot Noir, which is therefore used for a large part of the Blanc de Noirs produced in Germany.

Even after being developed into a red wine, however, it can be transformed into a white wine and thus into a Blanc de Noirs using infrared photography (Mangold et al., 2015). This takes advantage of the fact that the light of the near infrared range (780 nm to 3 μm) is scattered much less by the red wine than visible light (keyword "Rayleigh scattering", see info box). If our eyes were also sensitive to infrared light, we could easily see through red wine and it would appear clear to us like white wine.

The spectral sensitivity of the CCD chips used in commercial digital cameras ranges from about 400 to 1100 nm (Fig. 4.10). However, since the light of the near infrared range would make the image blurry and less contrasty and would worsen the color reproduction, infrared blocking filters

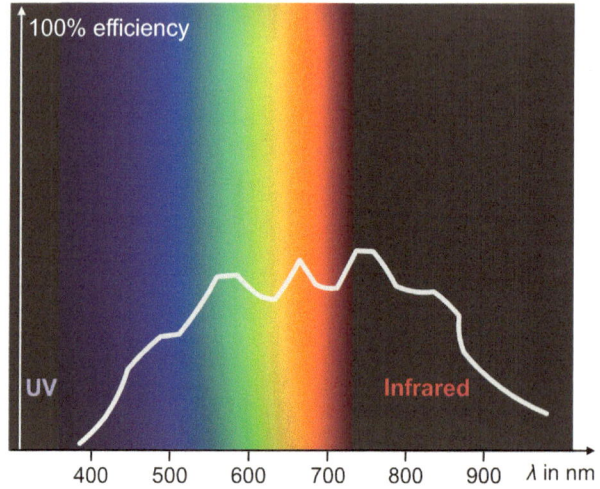

Fig. 4.10 Sensitivity curve of commercial CCD chips

are used in digital cameras, which suppress wavelengths above 700 nm (Fig. 4.11a). If you remove this filter from the beam path of a discarded digital camera and additionally use an infrared pass filter (Fig. 4.11b), which is held or screwed onto the lens, you get a cheap way to do digital infrared photography.

An example image of a deep dark red wine, taken with an infrared pass filter from smardy (cut-off wavelength 720 nm), is shown in Fig. 4.12c. The effect described above is clearly visible and the red wine becomes a Blanc de Noirs!

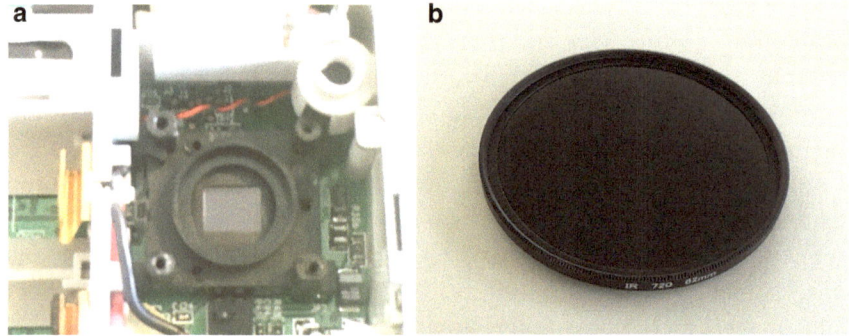

Fig. 4.11 Digital camera with IR filter (**a**), Infrared pass filter for screwing onto the camera objective (**b**)

Fig. 4.12 Red wine, taken with visible light (**a**), with infrared and visible light (**b**), only with infrared light (**c**)

Rayleigh scattering

The scattering named after *John William Strutt* or *Lord Rayleigh* (1842–1919) occurs when the wavelength λ of the scattered light is significantly larger than the size of the scattering particles. The intensity I of the scattered light is then inversely proportional to λ^4. Short-wave blue light (approx. 450 nm) is thus, for example, scattered more strongly by the air molecules of the atmosphere than long-wave red (approx. 650 nm):

$$\frac{I_{blue}}{I_{red}} = \frac{\lambda_{red}^4}{\lambda_{blue}^4} = \left(\frac{650\,nm}{450\,nm}\right)^4 \approx 4{,}4$$

Conversely, the so-called extinction, i.e., the significantly lesser attenuation of red light than of blue when passing through the atmosphere, is therefore the case. This is, among other things, the reason why the sky appears blue to us during the day and the setting sun appears red (Fig. 4.13). Due to the long path of light through the atmosphere at sunset, only the red component can reach us directly. The situation is completely analogous with our "Blanc de Noirs", which the infrared light passes with low scattering.

Fig. 4.13 Sunset viewed through a wine glass

5

Of Good and Bad Drops: Fluid Dynamics of Wine

In the next experiments, we want to deal with the behavior of wine as a liquid. The scientific branch of physics responsible for this is fluid dynamics or fluid mechanics. This sub-discipline helps us understand why pouring is a complicated matter or why even on the most beautiful evenings, wine tends to drip. We will also see that hydrophobia is not a disease, but leads to more cleanliness in nature. Finally, we want to send you to fetch wine—but please with a sieve!

Pouring? Yes Please, But Without Mishap!

Experiment 27: Spilling as a Law of Nature

The guests have taken their seats, the mood could not be more festive, all eyes are on the perfectly aerated wine in the host's hand. Barrique, as profoundly red as the tablecloth is innocently white. The bottle neck leans towards the glass and releases the precious content. Unfortunately, not only into the glass—the tablecloth also gets its share again. How clumsy! Was that necessary?

Yes, it was! Let's take a closer look (Fig. 5.1).

It seems as if spilling is a law of nature. Held by the glass of the bottle, the wine makes its way around the edge of the opening, undeterred in the direction of the outer bottle neck, and cannot be stopped anymore.

© The Author(s), under exclusive license to Springer-Verlag GmbH, DE, part of Springer Nature 2024
L. Kasper and P. Vogt, *Uncorking the Physics of Wine*,
https://doi.org/10.1007/978-3-662-68759-8_5

Fig. 5.1 Still image series of pouring from a wine bottle

The world of physics has long been aware of this problem. As early as the 1950s, it got its name: *teapot effect*. As far as spilling is concerned, wine and tea drinkers are obviously fellow sufferers. Since then, there have been and continue to be a number of attempts to explain it. Flow vortices in the jet of the poured liquid are supposed to make it cling to the outer wall of the spout. Physicists also thought they recognized the main culprit in the atmospheric air pressure. Even the *Ig-Nobel Prize*[1] was awarded in 1999 for a publication on the *teapot effect*.

The fact that the explanation of such an everyday phenomenon is not trivial is shown by ever new theoretical and experimental works. Ultimately, the process of separating the flow of liquid or the "flowing around" the edge depends on the flow speed, the inertia of the liquid, and the interaction forces between the liquid and the bottle neck or spout. These are described with the *hydrophilicity* of the wetted material, which means "water-loving". In the case of wine bottles, we are dealing with glass, which is such a hydrophilic material. A property that thus contributes to spilling.

Finally, the geometry of the lower edge of a spout also plays a crucial role. The smaller the radius of curvature of the edge over which the wine flows, the better the chance that the inertia forces of the flowing liquid will prevail, it can separate from the edge of the spout and thus not miss its target, the wine glass. However, with wine bottles, the edge of the bottle neck is rather not sufficiently sharp-edged—another potential excuse for possible mishaps.

---

[1] The *Ig-Nobel* is—literally translated as "unworthy prize"—actually a very coveted award among scientists, first awarded in 1991 at MIT (Massachusetts Institute of Technology).

Hydrophilic or hydrophobic?

The common interface of two substances that are in contact with each other is subject to the so-called *surface tension.* Depending on the sign of this tension, when a liquid comes into contact with a solid wall (e.g., water and glass), it leads to wetting or not. At the point of contact, a corresponding angle forms (Fig. 5.2a). Surfaces of solid substances that have a contact angle with an adjacent water surface of less than 90° are referred to as *hydrophilic.*

A material is referred to as *hydrophobic* (meaning "water-repelling" and thus the opposite of *hydrophilic*) if its surface has a contact angle with water of more than 90° (Fig. 5.2b). Water then tends to contract into a spherical shape. Such materials are found in nature in plants and serve, for example, for the cleaning of leaf surfaces. In technology, the lotus effect attributable to the plant is exploited, for example, in Teflon coatings.

The surface tension is due to the intermolecular forces of the liquid molecules among themselves and to the molecular interactions with the wall material. Therefore, this tension and thus also the wetting ability depend on the type of wall material and the liquid. The combination of water and glass usually results in a wetting situation. However, coatings intentionally or unintentionally applied to the glass can lead to hydrophobic effects.

The surface tension has the unit of a force per length: 1 N/m.

Since we do not want to change the type of liquid (wine) and the bottle material (glass) when pouring, all experimental evidence and theoretical considerations from hydromechanics for everyday life can be summarized into two simple action recommendations for pouring wine without mishap:

- Be brave and pour quickly! The inertia of the flowing liquid then "wins" more likely against the interaction forces with the glass of the bottle.
- Use "sharp-edged" pourers! It is then harder for the liquid to "flow around" the edge.

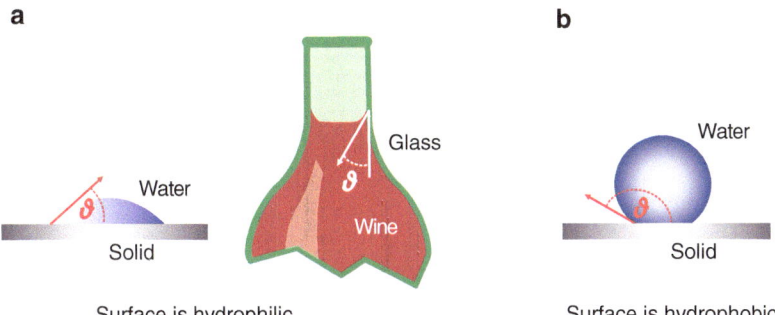

Fig. 5.2 Contact angle ϑ between a liquid and a solid; hydrophilic (**a**), hydrophobic (**b**)

Experiment 28: Carrying Wine with a Sieve?

Admittedly, there is hardly a good reason to try this. It would be a shame about the wine and it could damage one's reputation. After all, we know sayings like: *Talking to a fool is like carrying water in a sieve.* This suggests the futility of the efforts and common sense would like to agree here. But hold on, maybe the proverb is judging prematurely at this point! Let's approach the situation physically.

Many are already familiar with the following small experiment from school: We take a glass—in our case, of course, a wine glass—fill it to the brim with water and cover it, for example, with a beer mat. Now the whole thing is carefully turned upside down, with the beer mat being pressed firmly onto the glass opening when turning. Once the glass is upside down, the beer mat can be released. The result is—even if already known—always astonishing (Fig. 5.3a). The beer mat holds!

But what actually holds it? It is the pressure of the surrounding air. This acts independently of direction and orientation on all surfaces, including the beer mat, pressing it against the rim of the glass. On the other side of the beer mat, inside the glass, we also find a pressure. This, in the immediate vicinity of the beer mat, also almost corresponds to the external air pressure. (A more detailed discussion of the pressure in the glass is provided in the info box.) The liquid in the glass, therefore, does not "weigh" on the beer mat. The fact that the beer mat seal holds tight can be explained with the *adhesive force* between the molecules of the liquid and the beer mat, as well as with the *cohesive force* of the liquid molecules among themselves.

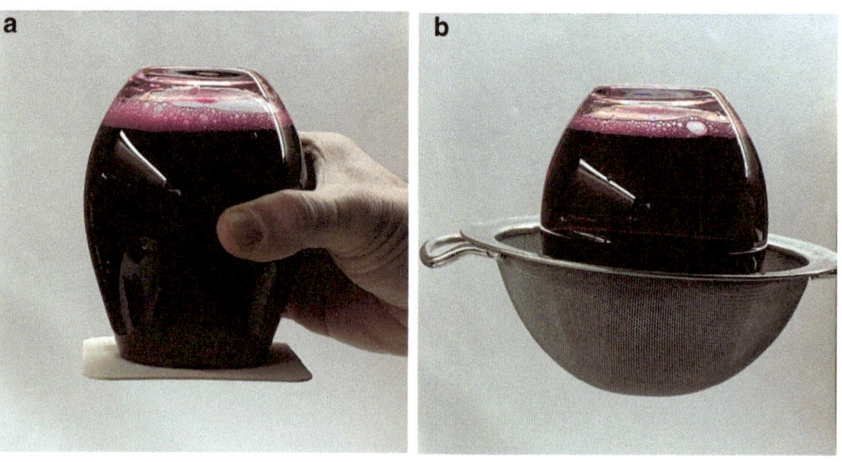

Fig. 5.3 The inverted glass holds tight! With a beer mat (**a**), with a sieve (**b**)

If the external air pressure and the pressure of the liquid at the very bottom of the inverted glass are equal, isn't the beer mat actually superfluous? As we know, the attempt without a supporting surface under the glass opening does not succeed. The smallest unevenness at the large interface of the liquid itself and at the adjacent contact line with the glass lead to deformations and these in turn to the "tearing" of the surface and finally to the leakage of the liquid. The beer mat, therefore, merely serves to stabilize the liquid surface.

Pressure Conditions in the "Beer Mat Experiment"

The *hydrostatic pressure* $p(h)$ of a liquid in an upright and open glass consists of the weight force of the liquid column per area and the *atmospheric pressure* p_0:

$$p(h) = \rho \cdot g \cdot h + p_0$$

Here, ρ is the density of the liquid, h is the height of the liquid column, and g is the gravitational acceleration 9.81 m/s^2.

From the equation, it can be inferred that the pressure increases linearly with depth and is maximum at the bottom of the glass. It decreases upwards and at the upper end—just below the level—the pressure is almost equal to the external pressure p_0.

For the inverted glass, the situation changes in that it is closed by the beer mat and the water. The following applies to the pressure in the liquid in this case:

$$p_{inside} = p_0 - \rho \cdot g \cdot h$$

Here, h again indicates the height of the liquid level in the glass. So now it applies that the pressure in the liquid at the bottom (at $h = 0$) in the immediate vicinity of the beer mat corresponds to the external air pressure p_0. Here too, the pressure decreases linearly with increasing height in the glass.

So the glass can indeed have a greater filling height, the pressure at the bottom of the beer mat remains at about the external pressure. The experiment can be carried out with a bottle filled to the brim, the opening of which is covered with a piece of rigid foil. That also holds tight!

How high could the glass (theoretically) be?

A theoretical limit for the maximum height of the glass is given by the fact that at the upper level the pressure becomes so low that the vapor pressure of the liquid is reached. At room temperature, the vapor pressure of water is about 25 hPa.

If you rearrange the equation given above for the inverted glass, and set the standard value for the atmospheric pressure p_0 and the vapor pressure p_{Vapor} for water, you get the limit height h_{max}:

$$h_{max} = \frac{p_0 - p_{vapor}}{\rho_{water} \cdot g} = \frac{(1013 - 25)\,\text{hPa}}{1000\,\text{kg} \cdot \text{m}^{-3} \cdot 9.81\,\text{m} \cdot \text{s}^{-2}} \approx 10.1\,\text{m}$$

This maximum height is, however, practically no longer feasible in this "beer mat closure" experiment.

Now we can vary the already astonishing beer mat experiment. Instead of a beer mat, we now use a sieve as the cover for the filled glass! This is placed on the still upright and filled glass. A not too large wine glass favors the success of the experiment, because this time the supporting hand must seal the surface of the glass opening covered by the sieve as watertight as possible when inverting the glass. If the glass is then in the "head-over position", nothing should leak out through the sealing palm. Then, carefully grab the sieve by its handle with the other hand and gently remove the hand under the sieve. With a bit of luck and skill, the water (or wine) remains in the glass and does not run through the sieve (Fig. 5.3b).

This is even more astonishing than the beer mat experiment! How does the sieve manage to hold the liquid? The sieve fulfills the same function as the beer mat did before. It stabilizes the liquid surface. However, it doesn't need to fully "support" this surface. A sufficiently dense network of sieve meshes is apparently enough, despite the openings in it. If the mesh size is small enough, then the surface tension of the liquid is sufficient to stabilize itself there. This also helps us understand how the entire glass opening covered by the sieve works.

By the way, this experiment can be taken even further—literally to the extreme: If you pierce a sieve mesh from below with a needle so that it protrudes into the glass, the liquid still does not leak out! Here again, the surface tension of the now even narrower spaces is the stabilizing factor.

Surface tension

Attractive forces between the molecules of a liquid cause the surface of the liquid to be comparable to a membrane under mechanical tension. This force per length, referred to as surface tension, acts parallel to the surface. The *surface tension* σ corresponds to the *surface energy* W_A per area:

$$\sigma = \frac{W_A}{A}$$

A liquid surface—as a "tense skin"—has a minimum energy at a given surface tension when the area also becomes minimal, and therefore strives to contract into a spherical shape. An example of this is the formation of drops.

Experiment 29: Tears of Wine

A wonderful red wine is in the glass, and that's certainly no reason to cry! And yet, tears can be involved here. Try it yourself out. Swirl a filled glass as evenly as possible all around, so that the wine wets a larger glass surface on the inside. Now put the glass down and wait about 10 seconds, then you can recognize these typical streak patterns as in Fig. 5.4. The streaks eventually form tears that run down in a narrow channel in the glass. Try varying the viewing angle or lighting to better perceive the phenomenon. As long as the glass is not drunk, the experiment is repeatable.

By the way, this phenomenon has long been known in science. The Brit *James Thomson* (1822–1892) first described it in 1855 as "curious motions observable at the surfaces of wine …". 16 years later, the Italian *Carlo Marangoni* (1840–1925) published a physical explanation of the effect, which was later named after him.

How do these patterns and tears actually occur? A first clue is provided by the almost trivial observation that it is not possible to produce these streaks and tears in the glass with water. It requires the mixture of water and alcohol. By the way, spirits also show this behavior.

After swirling the wine in the glass, a liquid film remains adhered to the inner wall, which has a large surface area (Fig. 5.5a). On this surface, especially in the upper area, the alcohol evaporates faster than the water. As a result, the proportion of water in the thin layer increases (Fig. 5.5b). Because water has a significantly higher surface tension than ethanol, the surface tension in the remaining liquid film increases and it tries to "contract" (Fig. 5.5c). This even results in a movement that is counter to gravity. At the upper edge of the liquid film, a wave crest forms, and the liquid also forms

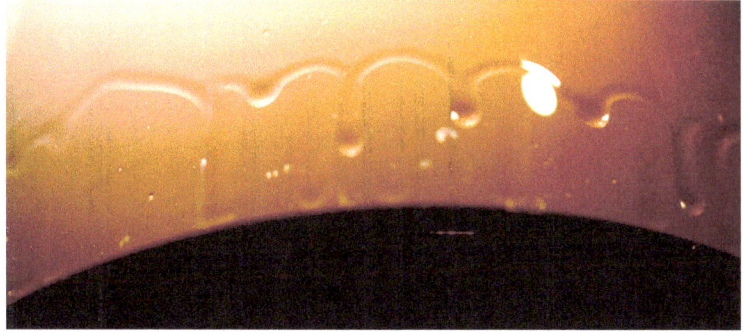

Fig. 5.4   Red wine forms typical "tears" in the glass after swirling

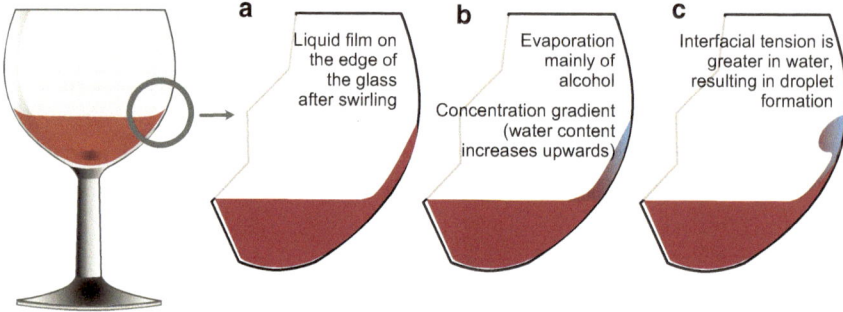

Fig. 5.5 Schematic explanation of the Marangoni effect in wine glasses (a–c)

drainage channels in a vertical direction, down which it then runs in tear-like drops.

Surface tension and Marangoni convection

The *Marangoni effect* can be explained by a gradient of surface tension σ of a liquid. This gradient can be caused thermally, but this is hardly relevant for the case of "wine tears". On the other hand, the gradient of surface tension can also be based on a concentration gradient of solutions.

In wine, ethanol and water are dissolved in each other. The alcohol content is between 10 and 15 volume percent. Due to the faster evaporation of the alcohol in the upper areas of the liquid film, a concentration gradient occurs, which drives the Marangoni convection, an upward flow against gravity. The strength of this concentration-induced convection can be described by the *Marangoni number* Ma, which is calculated as follows (Sun, 2018):

$$\mathrm{Ma} = -\frac{\partial \sigma}{\partial c} \frac{\Delta c \cdot L}{\eta \cdot \kappa}$$

The relationship shows that the convection is proportionally larger, the larger the gradient of the interfacial tension $\frac{\partial \sigma}{\partial c}$ is in relation to the concentration, the larger the concentration change Δc is, and the smaller the dynamic viscosity η and the thermal diffusivity κ are. L is the characteristic length dimension of the liquid.

6

The Well-Tempered Wine

The next experiments are three typical cases for thermodynamics. At the beginning, we ask about the correct temperature of a wine. It turns out that the greater challenge is to keep a wine at an optimal lower temperature level in a warm environment than vice versa. For this, we present a simple and effective tool. Of course, one could also cool with ice cubes. With wine, this is not well received among connoisseurs, but for cocktails, it is part of the process. On this occasion, we discover that ice cubes sometimes float and sometimes sink and draw our conclusions from this. And while we're on the subject of cocktails: We naturally know a brandy and also know how it was made by distilling. But the "burn" can also be achieved without heat, namely at really low temperatures.

Frapper or Chambrer?

Do you know not? Don't worry, this is clearly insider knowledge. Under *frapper*, especially the French wine friends understand the rapid cooling of a too warm white wine. Whereas one tries to warm up an undercooled red wine through *chambrer*. After all, it seems to be simple: white wine belongs in the refrigerator and red wine is served at room temperature. Unfortunately, this is not quite true and it is not that simple! On the one hand, white and red wines need to be differentiated, and on the other hand, there are also sparkling wines and other "specialities" such as ice wines. The following will deal with the correct serving or drinking temperature. It is

L. Kasper and P. Vogt, *Uncorking the Physics of Wine*, https://doi.org/10.1007/978-3-662-68759-8_6

clear that the storage temperature has an extraordinary influence on the maturation and the later aromas of the wine. But we usually don't have this in our hands and have to rely on the wineries and dealers we trust. However, we do have it in our hands to serve our guests (and ourselves) a well-tempered wine. A quick orientation is provided by the following overview of the drinking temperatures recommended by wine friends for different types of wine (Table 6.1).

However, this overview can only provide a first orientation. Here, above all, the individual palate is in demand and after all, they are not such bad "apprenticeship years" in which one can gain one's own experiences.

Experiment 30: "Eco-Refrigerator" for a Wine Bottle

Let's assume the following situation: A summery picnic in nature is planned and a light fruity white wine should definitely be included. According to the recommendation in Table 6.1, the wine should then best be enjoyed at a temperature of 8 to 10 °C. Before setting off, the bottle is taken from the refrigerator at about 6 °C. Of course, we are happy about real, i.e. warm summer days, but the wine then suffers.

A simple designed wine cooler provides relief here. It is made of pure clay and—this is important—is not glazed! Because if it were glazed, it might look a bit more interesting, but it would only be half as good for its actual purpose.

The trick is that the cooler is given an extensive water bath before use. It should soak up plenty of water. When it is then in the fresh air, the water begins to evaporate on the surface of the cooler. A light breeze enhances the effect. Physically, evaporation means the release of energy to the surroundings of the surface (see *heat of evaporation* in the info box). This energy loss is noticeable as cooling of the body that gives off the water, i.e., the walls of

Table 6.1 Optimal drinking temperatures of different types of wine (see www.wein-freunde.de)

Sparkling wines (Prosecco, simple Champagne)	5–7 °C
Vintage champagne	8–10 °C
Light white wines (e.g. Riesling)	8–10 °C
Full-bodied white sweet wines (e.g. Ice wine)	8–12 °C
More full-bodied white wines (e.g. Pinot Gris)	10–12 °C
Full-bodied white wines (e.g. barrel-aged Chardonnay)	12–14 °C
Light red wines (e.g. Beaujolais)	12–14 °C
Medium-bodied red wines (e.g. Chianti Classico)	14–17 °C
Tannin-rich and older red wines (e.g. Barolo)	15–18 °C

the clay vessel. You can quickly feel that the cooling "kicks in" when you put your hand on the clay surface. This also keeps the air between the bottle and the clay wall colder and thus reduces the heat output of the bottle. The more water such a clay cooler has absorbed, the longer its cooling function lasts. It is also clear that the decoratively glazed wine cooler does not make a good impression here.

What effect can one expect from such a cooler? In an experiment, we compared the warming behavior of two identical bottles with the same fill quantity and starting temperature. The setup of the measurement is shown in Fig. 6.1. One of the bottles is in a previously watered clay cooler, the other is "naked". Both bottles contain 0.75 l of refrigerator temperature water (6 °C). A digital thermometer is inserted in both bottles, with the probes each protruding into the middle of the bottles. The ambient temperature in the laboratory was 22 °C.

We let the measurement run for 90 min. The result is illustrated in Fig. 6.2. Even if the effect does not look outstanding, it is the case that the temperature range of 8 to 10 °C, which is recommended as optimal, was already left in the uncooled bottle after 30 min, while the bottle in the cooler remains well-tempered for twice as long. The 12-degree mark is reached

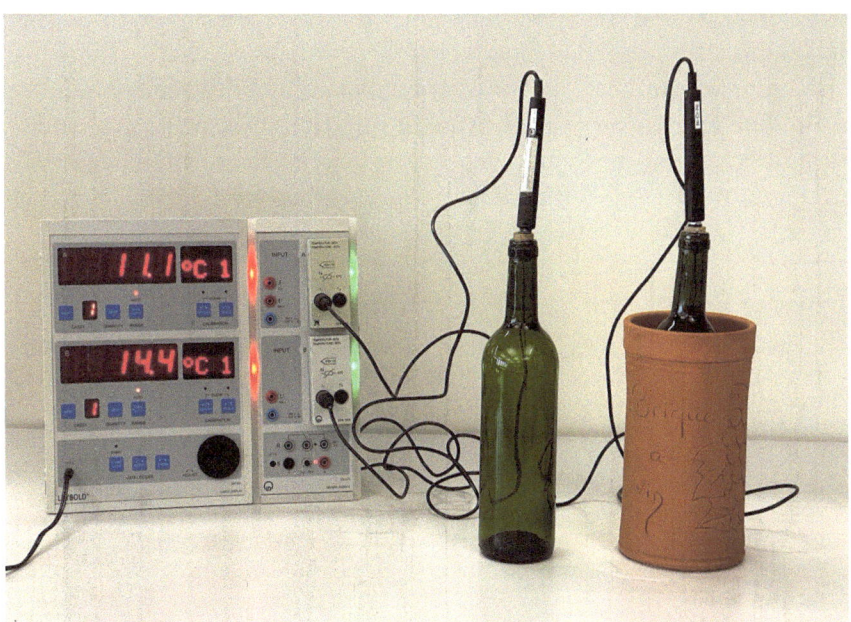

Fig. 6.1 Comparative measurement of the effectiveness of a simple clay wine cooler

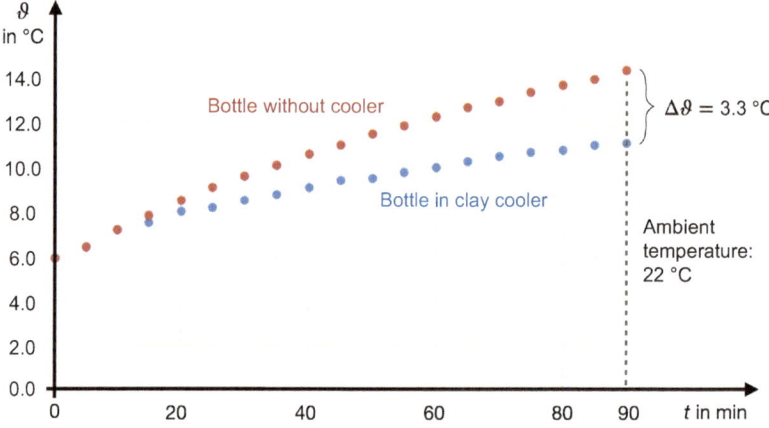

Fig. 6.2 Result of the measurement: The wine cooler does its job!

without cooling after one hour, with cooling it is not exceeded at all within the measurement time.

Use outdoors with perhaps a bit of wind and at a higher outside temperature would even improve the laboratory result. And this type of cooling is ecological anyway. By the way, you can quickly reproduce this experimentally on your terrace in summer. Even if your refrigerator should fail, it is best to take a large clay flower pot and water it generously. With a cover on it, you can keep perishable food longer than without cooling.

We humans also regulate our body temperature in this way when it begins to rise due to exertion or in the case of infections. We then sweat and with the evaporation of the liquid, we release energy to the surroundings and as a result, our body temperature drops.

Heat of Vaporization

In the molecular model, a liquid consists of a compound of molecules that can be easily shifted against each other, but are bound together by attractive forces (cohesive forces). During evaporation, individual molecules leave this compound and must do work against the attractive forces. Only the fastest molecules have the necessary energy for this. This results in a reduction of the average molecular speed and thus the temperature for the remaining molecules. As a result, the liquid cools down. This energy loss is material-dependent. If it is related to one kilogram of the liquid, it is called *specific heat of vaporization* Q_V. For water, for example, the specific heat of vaporization is 2253 kJ/kg (at the boiling point $T_S = 373.2$ K), for ethanol it is 844 kJ/kg (at the boiling point $T_S = 351.6$ K).

The energy withdrawn from the liquid by evaporation of a mass m is thus given by:

$$\Delta E = Q_V \cdot m$$

In contrast to *vaporization at the boiling point*, during *evaporation* the transition from the liquid to the gaseous state occurs below the boiling point of a liquid.

Experiment 31: Wine Cocktail on the Rocks or: Is the Hugo Overflowing Now?

How strict are you on the topic of "ice cubes in wine"? For some, it's a natural need in summer, while others might call it "heresy". As in many cases, we advise against black-and-white thinking. After all, if it tastes good, it should be allowed. At this point, a tip can be given: Keep a few wine berries in the freezer in summer instead of ice cubes to cool the wine.

Undisputedly more popular, however, is the use of ice cubes in wine cocktails. The Spanish *Sangria* or the *Hugo* popular in German-speaking countries are popular representatives. A simple basic recipe for the latter cocktail is:

15 cl Prosecco, 2 cl lemon balm syrup, a splash of soda and ice cubes stirred in a wine glass; optional: a sprig of mint, a slice of lemon.

Now you put all your effort into this recipe or a creative development of your own and serve your guests the cocktails. Not only did you mean well when mixing, but also when pouring. The glasses are filled to the brim and in addition, the ice cubes protrude significantly above the level. If the conversation at the table is so exciting that the cocktails are left standing for a while, the ice will inevitably melt. And then? Will the Hugo overflow?

Let's take a closer look at the process under laboratory conditions before the next party. Water can suffice as a "cocktail model" for us, it makes up about 90% of the real cocktail anyway. We will exaggerate a bit with the ice to see the effects clearly. The melting process in this experiment is shown in its temporal development in Fig. 6.3.

The glass shown is exactly filled to the brim with water, in addition, the initially quite large lump of ice protrudes above the rim of the glass by a few millimeters. We see that—although more and more ice melts—the water level in the glass remains constant. Even after the complete melting, the water level still exactly matches the rim of the glass. In a serious case,

Fig. 6.3 Course of the melting process of an overhanging lump of ice in a glass filled to the brim

the tablecloth at the party would be saved, provided the guests have a steady hand. But how does this constant fill level actually come about?

For the explanation, we need the Archimedean principle (see Experiment 46 in the info box): The buoyant force on a body immersed in a liquid (or in a gas) corresponds to the weight force of the liquid (or gas) displaced by this body. In our case, the body is a block of ice that is immersed in water. The fact that ice floats on liquid water is due to its slightly lower density (more on this in the info box), i.e., a certain volume of water has a higher weight than the same volume of ice. Therefore, ice floats on the surface, but does not fully immerse, only about 90%. When the ice has completely melted, its meltwater has the same density as the existing water and thus occupies exactly the volume that was previously displaced by the ice cube. Consequently, the liquid level remains the same.

Density anomaly of water

With most substances known to us, the density increases with decreasing temperature, they "contract" when cooled. Not so with water. It has its greatest density at a temperature of 4 °C. This means that when cooling from a higher temperature to 4 °C, water behaves "normally". However, with further cooling, it expands again. (Therefore: Don't put beer in the freezer!)

Because this behavior of water deviates from that of most substances, it is referred to as *density anomaly of water*. In the molecular picture, this can be explained by the continuous temperature-dependent rearrangement of the water molecules. At a temperature of 4 °C, the water molecules linked via hydrogen bonds occupy the smallest volume.

Since we are talking about cocktails with ice here: In a Hugo, the ice floats on the surface in the glass. Caution is advised if you order a cocktail and then see that the ice has sunk to the bottom of the glass. When does this happen?

One idea would be heavy water. Ice cubes made from this water would also sink in a non-alcoholic cocktail. But we don't want to drink that! If we disregard the influence of dissolved sugar here for once, it must therefore apply that the density of the ice is greater than the average density of the cocktail, which consists essentially of the components water and ethanol: $\rho_{ice} > \rho_{cocktail}$. Let the densities be

$$\rho_{ice} = 0.918 \ \frac{g}{cm^3}, \ \rho_{ethanol} = 0.789 \ \frac{g}{cm^3}, \text{and} \ \rho_{water} = 0.998 \ \frac{g}{cm^3}.$$

Obviously, the cocktail is so strongly mixed that its density is less than that of the ice cubes. How much volume percent alcohol must the cocktail have at least for this? For an estimate, we make the following approach:

$$\rho_{ice} > \rho_{cocktail} = x \cdot \rho_{ethanol} + (1 - x) \cdot \rho_{water}$$

(x: proportion of ethanol, which is between 0 and 1). Solving for x and inserting the numerical values leads to an alcohol content of the cocktail of at least 42%. If your ice cubes ever sink to the bottom, you have, strictly speaking, not a cocktail in your glass, but most likely pure spirits (Fig. 6.4)!

Fig. 6.4 Ice cube sinks in rum with 80% Vol. (**a**), ice cube floats in grain brandy with 28% Vol. (**b**)

A Digestif? Even when Frozen!

Experiment 32: "Freezer Burn"

After the wine cocktail on the rocks, this experiment gets a bit colder and also high-proof. After dinner, it's time to stimulate digestion, i.e., it's time for a digestif! To stay on the topic of "wine", we suggest a wine or pomace brandy for this—e.g. a Cognac, an Armagnac or a Grappa. As the nickname "brandy" clearly indicates, these digestifs or schnapps in general are obtained by distillation. This involves heating an alcoholic starting product, usually a mash set with dry yeast several weeks before. The process is based on the different boiling temperatures of alcohol (for ethanol 78 °C) and water (100 °C). Due to the lower boiling point of alcohol, it evaporates earlier—the steam is fed to a condenser, cooled and liquefied. Usually, one or even two more distillations are carried out with the alcohol obtained in this way.

The separation of alcohol and water can not only be achieved by heating, but also by cooling down the starting product. Yes, you read correctly, you can actually make your own schnapps with your freezer and according to our research there is no law that prohibits you from "freezing schnapps"—however, we cannot guarantee that the tax office will not knock on your door one day!

The procedure is incredibly simple and has already been described in a similar way in the book *Incredibly Simple. Simply Incredible: Physics for Every Day* by *Werner Gruber*: You pour a bottle of wine into a PET bottle and put it in the freezer (Fig. 6.5). When freezing, the wine expands slightly, which is why the bottle should not be completely filled. Since the freezing point of ethanol is at −114 °C and thus far below the freezer temperature (approx. −18 °C), the water freezes completely, but a small part of the ethanol remains liquid and can be decanted the next day. It is best to place the bottle open and upside down in a container and thaw about a fifth of the starting quantity in addition (Fig. 6.6).

Fig. 6.7 shows a measurement of the alcohol content of the original wine (14% Vol.) and the obtained "freeze distillate" (20.5% Vol.). Admittedly, it's not yet a proper schnapps with 40% Vol., but the effect is clearly visible and as with actual schnapps distillation, you could repeat the process with the obtained "grape liqueur".

After numerous tests, however, we must admit that the achievable quality could not convince even after various refinement attempts. So we would like to advise you to continue to stick with your preferred wine or pomace brandy.

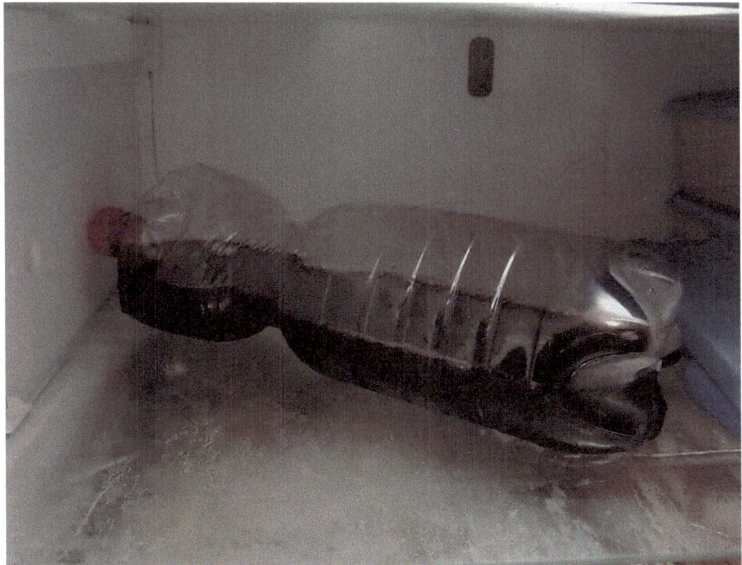

Fig. 6.5   PET bottle filled with wine in freezer

Fig. 6.6   Decanting the ethanol and thawing part of the frozen wine

From a physical point of view, the "self-frozen" is interesting and if you want to test it, please pay attention to the following note: Since the various alcohols are not separated from each other in the presented procedure, dangerous methanol would also end up in your schnapps if you used mash. Therefore, only use high-quality wine or beer for freezing and never mash!

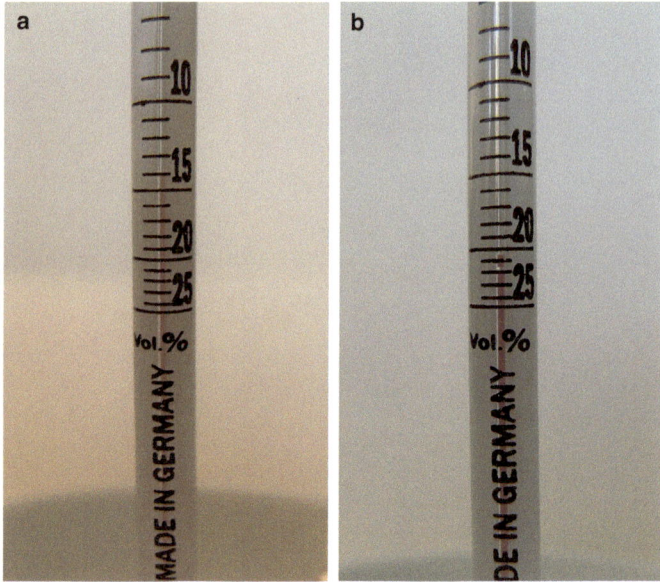

Fig. 6.7 Determination of the alcohol content before freezing (**a**) and after decanting (**b**) using a vinometer

Functioning of a Vinometer

To determine the alcohol content of a water-alcohol mixture, a vinometer can be used (Fig. 6.8). The measuring range usually extends from 0 to 25% Vol., with a measurement accuracy of ±0.5% Vol. For the measurement, some wine is poured into the funnel of the vinometer and it is ensured that the capillary tube is completely filled. If necessary, one can lightly blow into the funnel until one to two drops of the liquid to be tested leave the opposite side of the capillary tube. The measuring instrument is then turned over and set down.

Under the influence of the gravitational force acting on the liquid column, the level continuously decreases. If the rise height no longer changes, the alcohol content of the wine can be read off the printed scale.

The functioning of the vinometer is based on the one hand on the surface tension of the wine due to acting cohesive forces and on the other hand on the interfacial tension between wine and glass due to adhesion (see info box of experiment 39). Both together lead to the fact that the liquid column does not sink further at the so-called *capillary rise height h*. With the surface tension σ, the contact angle θ (Fig. 6.9), the liquid density ρ, the gravitational acceleration g and the tube radius r, the capillary rise height can be estimated with the following relationship:

$$h = \frac{2\sigma\cos\theta}{\rho gr}$$

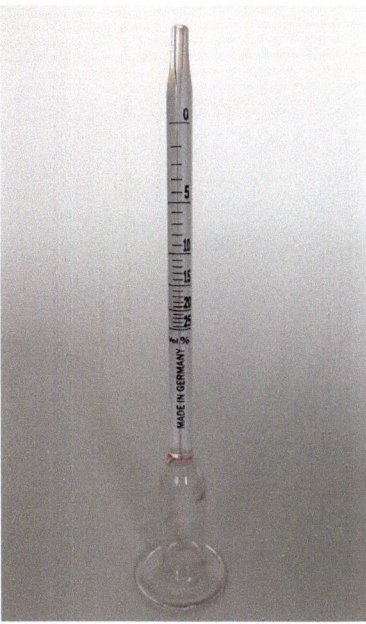

Fig. 6.8   Determination of the alcohol content using a vinometer

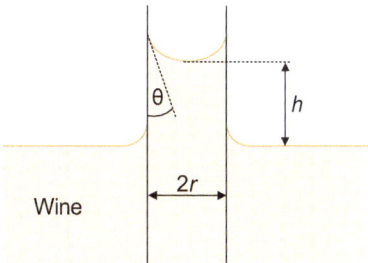

Fig. 6.9   Capillary rise height and contact angle

The determination of the alcohol content via the capillary rise height is possible because the surface tension of alcohol is smaller than that of water. The more alcohol the wine contains, the lower is thus its capillary rise height. However, the surface tension also depends on the sugar content of the liquid, which is why the vinometer only provides an accurate measurement result for dry wines without residual sweetness. The surface tension increases with the sugar content, i.e., for semi-dry and sweet wines, a vinometer measurement would yield too high an alcohol content.

7

Magic Tricks and Wonders: Acrobatic Mechanics

This chapter offers a colorful bouquet of physical ideas at the party table. However, we start with ancient inventions in the first experiments. These easily give the impression of a raised index finger and indeed remind us to exercise moderation. For this purpose, the old masters developed automatons that allow strict allocation from wine jugs or also regulated the mixing of wine and water. Accordingly, we introduce you to intelligent glasses that are capable of withdrawing all the wine already in the glass from those—and only those, who overdo it when pouring. An automaton that can turn water into wine then sounds significantly more conciliatory.

After our "ancient lesson" you can brace yourself! A series of experiments follows, which can best be summarized under the keyword "party tricks". In the physical sense, this involves sometimes spectacular applications of mechanics: from smaller "center of gravity magic" to demonstrations that will impress your guests next time.

Ancient Engineers—Water to Wine or Moderation?

None of us would like to experience this: dear guests have unexpectedly come to visit and the wine rack is empty! Then only a miraculous machine that turns water into wine can help. This machine actually exists and was invented by the Greek scholar *Heron of Alexandria* (probably 1st century), among many other curious and ingenious machines. In his works

© The Author(s), under exclusive license to Springer-Verlag GmbH, DE, part of Springer Nature 2024
L. Kasper and P. Vogt, *Uncorking the Physics of Wine*, https://doi.org/10.1007/978-3-662-68759-8_7

Pneumatika and *Automata*, Heron proves himself to be an excellent engineer. He is credited with the automatic opening of a temple gate after lighting a sacrificial fire and the earliest precursor of steam engines developed one and a half millennia later. Although *Ernst Mach* (1838–1916) denies the scientific nature of Heron's mainly crowd-pleasing inventions in his book *Science of Mechanics (translated 1919)*, the abundant contributions to mechanics and their extraordinary influence on the dissemination and further development of science cannot be denied.

It is quite obvious that *Heron* was also a friend of wine. A whole series of ingenious inventions are dedicated to dealing with the grape juice: the pouring of wine with various automatic filling quantity limits, the optional pouring of different types of wine or even water from the same vessel, the automatic mixing of water and wine in any ratios, the filling of wine glasses from a jug that never empties or—and this brings us back to the initial problem—the transformation of water into wine. This little miracle is achieved with one of the various wine automatons by *Heron*, which we will investigate in the following experiment.

Experiment 33: Heron's Wine Automaton

The automaton presented here is one of several variants that are attributed to *Heron of Alexandria*. It is capable of "transforming" one liquid into another. In our case, of course, we wish to obtain red wine from water. By the way, the automaton doesn't care, it can also do it the other way around …

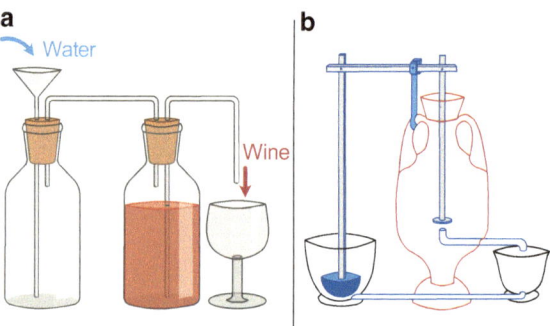

Fig. 7.1 Heron's wine automaton, which turns water into wine (**a**); a wine jug that automatically refills a vessel (**b**)

But how does it do it? Fig. 7.1a shows the principle of the wine automaton. If you want to turn water into wine, you prepare the automaton as shown. An empty container is directly connected to a second one, which is filled with the "target liquid"—here red wine. Furthermore, a tube is inserted into the cork of the empty vessel, which is used for filling and protrudes into the vessel. The water is poured into it.

A U-shaped tube also protrudes into the vessel filled with wine, the long leg of which reaches just above the bottom. The short leg of the tube forms the spout. It is important to ensure that all connections are airtight.

When water enters the empty vessel, it occupies a certain volume that the air had previously claimed there. Assuming airtight connections, the air can only escape via the connecting tube into the neighboring vessel. But there is already air and especially red wine there. The air itself cannot escape further. It is slightly compressed, but above all, the pressure increases. There is also pressure in the red wine. This consists of the *hydrostatic pressure* p_s of the liquid at a certain depth and the air pressure p_{air} above the level:

$$p_{liquid} = p_s + p_{air}$$

The key to the rise of the red wine in the capillary tube is the pressure at the location of the lower end of the tube. Therefore, it is important that it protrudes just above the bottom of the vessel.

As long as the air pressure above the wine level is equal to the atmospheric pressure, nothing happens. The wine in the capillary tube rises until it matches the level in the vessel.

The moment water is added to the other vessel, however, the air pressure in the entire two-bottle system increases. The air cannot escape back because the filling tube also extends deep down over the bottom of the vessel and the added water blocks the return path. Thus, the pressure in the red wine increases and drives the liquid in the capillary tube upwards. If more water is added, the wine can eventually overcome the maximum height and find its way through the spout into the wine glass.

As beautiful as the exploration of the Heron's wine automaton is, it unfortunately also brings us the realization that it is not the solution for the case of unexpected visitors. It would have been too good to be true. Due to the compressibility of the air, by the way, not exactly the same amount of the added water is "transformed" into wine—but almost.

Hydrostatic pressure in liquids

Hydrostatic pressure p_s in a liquid refers to the pressure at a certain depth, which results from the weight force of the liquid above it. Points at the same depth have the same hydrostatic pressure.

Since liquids are almost incompressible, it can be assumed that for the dimension of the vessel, the density ρ does not depend on the height. Then the simple expression for the hydrostatic pressure is obtained

$$p_s = \rho \cdot g \cdot \Delta h$$

Where g is the acceleration due to gravity and Δh is the depth (or the height of the water column above the point where the pressure is determined). The increase in hydrostatic pressure for water is exactly 1 bar at a depth of 10 m and 1 bar for every additional 10 m.

Finally, let's take a brief look at another "wine invention" by *Heron*. This is a device that automatically refills just as much wine in a vessel as was previously removed from it.

In Fig. 7.1b, we imagine the large amphora filled with wine. In the depicted situation, the two vessels on the left and right are still empty, but the inlet to the spout in the amphora is open. So, the right vessel is being filled, and so is the left one, as they are connected by a tube. As the vessels fill, the float in the left vessel rises. Through a lever mechanism, the inner closure of the inlet to the spout in the amphora then lowers. At a certain fill level of the vessels, the inlet is tightly closed.

If you now remove a certain amount of wine from the right vessel by skimming, the float on the left lowers and the closure in the amphora lifts. As a result, the spout lets just as much back into the vessels as was previously removed from them, until it is closed again.

Our next experiment also involves the automatic regulation of a certain amount of wine. However, it is even more moralizing than the invention of *Heron*…

Experiment 34: With Pythagoras to Moderation

How can a wine glass and moderation go together? With the drinking vessel that is now the focus of our attention, this connection is indeed possible. Not that the glass is particularly small. That would be too easy. Rather, physical acumen, a mischievous cunning, and a raised index finger are all combined in this vessel.

Appropriately, this hydrodynamic invention *Pythagoras of Samos* (around 570 to 480 BC) is attributed, who himself is said to have ascetic traits and a vegetarian diet. Although there is hardly any reliable knowledge about him as a person, many myths surround this master and founder of a school of thought. It is reported that *Pythagoras* was concerned about the table manners of some craftsmen he had commissioned and helped out with a peculiar cup. For this very purpose, he used hydrodynamic laws to construct a special mechanism in the cup, which we will look at more closely in the next experiment.

The vessel, named in honor of its inventor as *Pythagorean Cup*, had a nasty property. It behaved "normally" as long as one did not demand to fill it as much as possible. That is, at an appropriate fill level, the cup could be used as one would expect. But woe to anyone who desired more than his table companions and held the cup too long for pouring. Suddenly the cup seemed to be no longer tight. It poured over the lap of the puzzled thirsty person and only stopped when nothing was left in it. *Cup of Justice* is therefore also called this drinking vessel.

Another name that is circulating is also *Tantalus Cup*. This name also indicates punishment for misconduct. *Tantalos* (lat.: *Tantalus*) was punished very harshly, he suffered the hellish Tantalos torments after he committed theft at the divine table to which he was invited. But it got much worse. He put the omnipotence of the immortals to a horrific test when they were once his guests, by serving them his own youngest son in a concealed manner as a meal. Of course, the gods noticed the monstrous and made *Tantalos* pay for this crime just as cruelly. Banished to the deepest depths of Hades, *Tantalos* found himself standing in water and yet could not quench his thirst. Whenever he bent his head down, parched and thirsty, the water disappeared. Above him, however, hung the most wonderful fruit-laden branches, which moved to unreachable heights as soon as he began to reach for them.

Even though this only constituted a part of the punishment and curse of the gods for *Tantalos*, it can be summarized briefly: Wanted too much and lost everything! This brings us back to the far less harmful drinking vessel with its ceramic or glass warning: Fill me to the brim and you will not have a single drop left…

Pythagorean cups are available as tourist souvenirs on the island of Samos. Let's take a close look at such a cup. Although the cylindrical bulge protruding in the middle arouses suspicion, this ceramic cup does not reveal its interior. However, these cups are also offered by some glassblowing workshops and laboratory glass manufacturers and in this form provide insight into their construction (Fig. 7.2a).

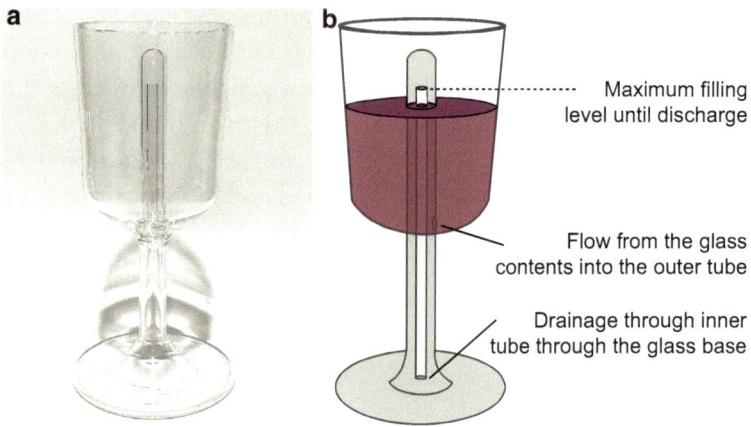

Fig. 7.2 Photo (**a**) and diagram (**b**) of a Tantalus or Pythagorean cup

The technical designs of the Tantalus cups can vary, but the basic principle is always the same. Let's look at the variant of the glass in the photo of Fig. 7.2a. The schematic representation next to it shows a double-walled glass tube in the axis of the glass as an extension of its stem. This tube has an opening on its outside near the bottom of the glass. When wine is poured in, the outer part of the tube fills with the same level as the contents of the wine glass. This happens in accordance with the so-called *principle of communicating tubes* in physics. The outer part of the tube is bounded at the bottom by the bottom of the glass. Its inner part itself represents another tube, which runs through the bottom of the glass and is open both at the top and at the foot of the glass.

When pouring the wine, the liquid levels in the glass and in the outer part of the tube rise in the same way. As long as the top end of the tube is not reached, the wine remains in the glass. However, after exceeding the critical height, the glass irrevocably leaks and almost completely empties. It is somewhat astonishing that the liquid in the tube apparently "runs uphill". How is this possible?

This surprising behavior can be explained with the principle of a *hydraulic siphon* (Fig. 7.3a). The flow in the tube is driven by the gravitational effect on the liquid. On the ISS space station, the Pythagorean cup would lose its educational effect! From the assumed gravitational effect, the potential energy of the liquid emerges as essential for understanding the siphon principle.

It is reassuring to know that during the outflow from the higher to the lower vessel, the center of gravity of the liquid considered as a whole

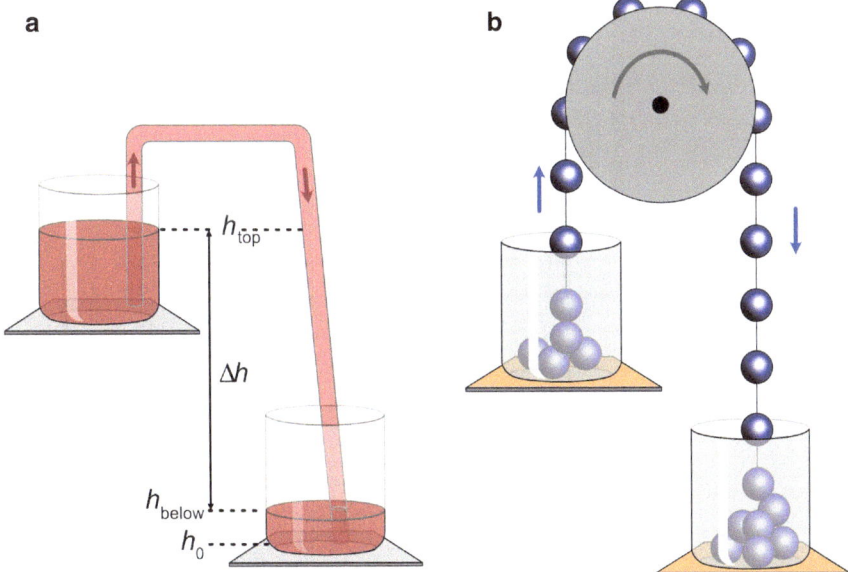

Fig. 7.3   Principle of the hydraulic siphon (**a**) and chain analogy (**b**)

decreases. The "uphill running" in the shorter pipe section of the siphon from the level height h_{top} to the maximum height of the pipe bend does not violate the law of energy conservation. The energy required to lift in this pipe section corresponds to the energy "gained" in the descending pipe section from the maximum height in the pipe bend back to the level height h_0. The decisive factor is therefore the pipe section in which the liquid passes through the height difference $\Delta h = h_{top} - h_{below}$. When the end of the long pipe dips into the liquid of the lower vessel, the gravitational pressure of the liquid must also be taken into account, which results from the rising level with the height $h_{below} - h_0$.

Just like the liquid in the siphon, a ball chain behaves as shown in the analogy in Fig. 7.3b. One is not surprised that the balls are not torn apart, they are held together by the thread. But if we turn back to the liquid, one might wonder why it does not tear apart in the upper pipe bend. After all, both liquid columns pull with their respective gravitational forces in opposite directions. So what is the "invisible thread" that maintains the cohesion of the liquid in the pipe? This results from the pressure at both ends of the pipe. Essentially, this is the external air pressure. This prevents the liquid from tearing and creating a vacuum or a vapor bubble at the "tear site".

The height of the pipe bend played no role in our considerations. For the Tantalus cup, this is also limited to the height of the cup. Nevertheless, it is a physically interesting question how high this arch could actually be constructed. The gravitational effects of the liquid in both pipe sections above the level height h_{top} cancel each other out. Mathematically speaking, it would not matter how high the pipe bend is. However, physics sets a limit here. When the value of the limiting air pressure is reached by a gravitational pressure $p_s = \rho g h_{max}$ of the liquid at a maximum height h_{max} of the pipe bend, then the liquid actually tears by forming vapor bubbles. At the earth's surface, this is about 10.3 m. Even with a strong pump, water in a pipe could not be sucked up beyond this height. It would then start to boil and form vapor bubbles.

Experiment 35: Thrifty by Diluting?

Let's now turn to a difficult topic: The German beverage called Schorle! Diluting wine may seem like a sin to some. And indeed, there are good wines whose bouquet would be irreversibly destroyed by adding water. On the other hand, the wine-and-water mixture has a long history. Not least, in antiquity, drinking diluted wine served as a precaution against diseases resulting from contaminated water due to the antibacterial effect of alcohol. We know this practice for sure from the Romans and Greeks. At least during the lifetime of *Heron of Alexandria*, around the first century, it was obviously common practice to enjoy wine diluted as well. This is evidenced by his inventions and devices, which were intended to automate this mixing. Perhaps *Heron* wanted to create a balance with another of his machines, which could turn water into wine (see Experiment 33).

And today? There's no question, Schorle is part of summer enjoyment and has cult status in some regions, e.g., in the Palatinate. The classic is a decent Riesling spritzer, but red wine can also be "spritzed". An important note when ordering the refreshing drink in different regions: Regardless of the fact that the German dictionary specifies the grammatical gender as "*die Schorle*" and less commonly also "*das Schorle*", in the Palatinate it is "*der Schorle*". There, both the use of a decent winemaker's wine and the correct mixing ratio are valued for the mixture. This limits the water content to a maximum of one third.

Now, with *Heron's* help, we can construct a wine-water mixing machine that we can even "program" to the mixing ratio of one third water and two thirds wine (Fig. 7.4).

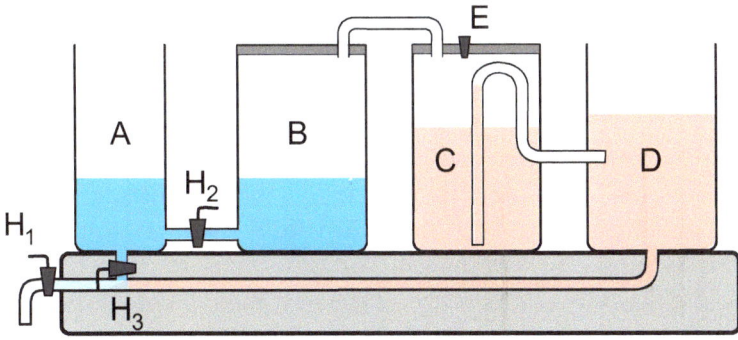

Fig. 7.4 Wine mixing machine with preset mixing ratio based on Heron (cf. Schmidt, 1899)

Heron's machine not only dispenses a certain mixture, but it also produces just as much of the mixture as you put water into the machine.

However, it must be noted here that the machine only produces the correct mixture with still water. Because this is sensitive to the pressure in the partly airtight containers, the intended ratio would be altered by the CO_2 gas of a sparkling water.

The desired mixing ratio is determined by the base areas of the two cylindrical containers A and B (Fig. 7.4). If the Palatinate minimum requirement of two thirds wine is to be met, then we have the ratio $V_{wine} : V_{water} = 2 : 1$. For the base areas of vessels A and B, it must then be: $A_A : A_B = 1 : 2$. If, for example, vessel A has a radius of $R_A = 10$ cm, then vessel B must be dimensioned so that its radius is $R_B = \sqrt{2}R_A \approx 14$ cm (this follows from $A = \pi R^2$). Another necessary preparation is to fill vessel C with the desired wine through the filling opening E and then close the opening again airtight. Finally, tap H_1 and tap H_3 must be closed and tap H_2 must be opened.

If, for example, one liter of the wine-water mixture is to be produced, one liter of water must be poured into the open vessel A. The tap H_2 at the connection of vessels A and B is open. Because of the physical principle of "communicating tubes" (see info box), two thirds of the water volume flow into vessel B, then both vessels have the same level. But since vessel B is airtight except for a connection to vessel C, the same volume, i.e., two thirds of a liter of air, escapes from vessel B into vessel C, which has already been filled with a supply of wine. Vessel C is also airtight. However, it has a siphon, whose vertical leg reaches almost to the bottom. The other leg leaves vessel C and ends in vessel D. The volume of wine displaced into vessel D by the siphon is the same as the volume of air previously pressed into vessel C, i.e., two thirds of a liter.

Now, in vessel A there is one third of a liter of water and in vessel D there are two thirds of a liter of wine ready for mixing. If valve H_2 is now closed, H_3 and H_1 can be opened and the liquids flow out of both vessels and mix in front of valve H_1. The result is one liter of the mixture in the desired ratio. If another mixture is to be created from the remaining wine supply, valve H_2 is opened beforehand, allowing the water to drain from vessels A and B. The machine is then ready for the next mixing order.

Communicating Tubes

If a system of interconnected tubes is filled with a liquid of density ρ, the liquid stands at the same height in all legs of the connected tubes. The shape of the tubes does not matter (Fig. 7.5)—a circumstance also referred to as the *hydrostatic paradox*. The fact that the pressure at the bottom of a column of liquid is independent of the cross-section results from the weight force of a column of liquid with the height h:

$$F_G = m \cdot g = \rho \cdot h \cdot A \cdot g$$

For the pressure p at the bottom of the vessel with the cross-section A, the following applies:

$$p = \frac{F_G}{A} = \rho \cdot h \cdot g$$

The *principle of communicating tubes* is widely used in everyday life. Examples include hose levels, water supply through water towers, or height compensation at river and canal locks.

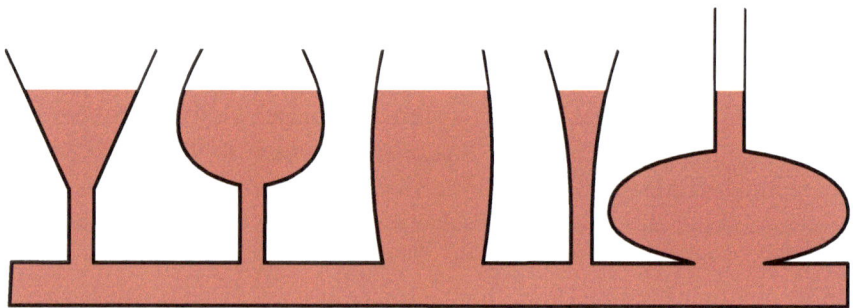

Fig. 7.5 Principle of communicating tubes

Experiment 36: The Wine Glass as a Diving Bell

We continue with the topic "Pressure in Liquids" and turn to a technical invention also attributed to antiquity: the diving bell. *Aristotle* (384–322 BC) already mentioned a turned-over kettle around 320 BC, which, for example, serves a pearl diver as a reservoir for breathing air and enables him to work longer. This was the birth of the idea of the diving bell, which is still in use today.

Without having to put ourselves in danger, we can experimentally study the principle of an open diving bell at the table, provided we have an empty wine glass and a larger waterproof container. And if we can also find a tea light or a napkin, then the small demonstration can begin.

The large container is filled with about 10 to 15 cm of water. We let something float on the water that should actually stay dry. In Fig. 7.6a, a napkin has been chosen for this purpose, which is stuck in a tea light holder. Now we take the wine glass, turn it upside down and put it over the floating napkin. With some force, the glass is slowly submerged further and further into the water until it sits on the bottom. In Fig. 7.6b, the wine glass has arrived at the bottom as an open diving bell, and the napkin has remained dry. Only a very small amount of water has penetrated into the glass at the bottom. It is easy to imagine that such a diving bell, if appropriately dimensioned, also offers people space and above all breathing air for some time.

From a physical perspective, the pressure and volume ratios in the diving bell are interesting. In Fig. 7.6b and c, it can be seen that a small amount of water penetrates the glass due to the hydrostatic pressure. How does this behave in real diving bells, which are used for work or tourist purposes

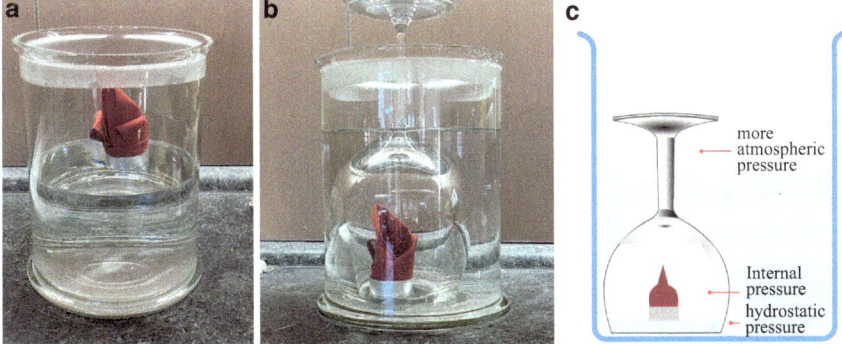

Fig. 7.6 Wine glass as a diving bell

Fig. 7.7 Zinnowitz diving gondola on peninsula Usedom (Baltic Sea); a dive with up to 24 people lasts 30 to 40 min

underwater (Fig. 7.7)? How high is the pressure in the bell and how much water penetrates?

To answer these questions, the laws resulting from the *general gas equation* (see info box) are needed. In this, the three essential quantities for gases *temperature*, *volume* and *pressure* are related to each other. If we assume for the situation of the diving bell that the temperature remains approximately constant, then the gas equation reduces to the statement: The product of pressure and volume of a closed gas quantity is constant (*Boyle-Mariotte law*). This law can now easily be applied to the diving bell. We must also consider that the same pressure prevails in the respective air volume of the diving bell as at the respective upper level under the bell. Furthermore, we need knowledge of the increase in the hydrostatic pressure of the water with depth. Here, the rough rule of thumb is that the hydrostatic pressure increases by 1 bar every 10 m. Figure 7.8 schematically shows these relationships. Starting from the atmospheric pressure above sea level (about 1 bar), 1 bar is added every 10 m depth. This pressure then also prevails in the gas volume in the diving bell. With the help of the *Boyle-Mariotte law*, the resulting gas volume in the diving bell can thus be determined.

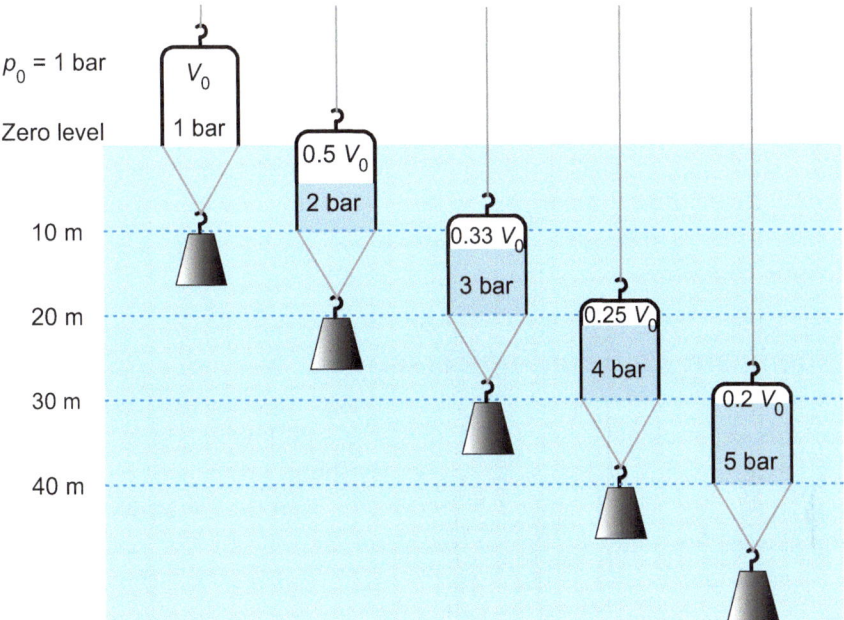

Fig. 7.8 Relationship of pressure and remaining air volume of a diving bell at different water depths

---

Isothermal State Change

The *general gas equation* describes the behavior of a gas when changing one or more of the quantities temperature T, pressure p and volume V:

$$\frac{p \cdot V}{T} = \text{constant}$$

For two states 1 and 2, the result is:

$$\frac{p_1 \cdot V_1}{T_1} = \frac{p_2 \cdot V_2}{T_2}$$

From the constancy of one of the three state quantities, three special cases can be derived. In the case that, as assumed for the open diving bell, for example, the temperature of a system remains constant (*isothermal state change*), the general gas equation becomes

$$p \cdot V = \text{constant or } p_1 \cdot V_1 = p_2 \cdot V_2.$$

This special case is named in honor of the physicists *Robert Boyle* (1627–1692) and *Edme Mariotte* (1620–1684) *Boyle-Mariotte Law*.

Balance Acts

The following experiments may not achieve any truly practical goals. Rather, they fall under the category of "party tricks" that will certainly attract attention. And yet, we can also benefit from the universal nature of physical laws here. Because what applies to the "balancing" utensils at the table also applies in many other practical life situations. Let's start the games with two floating corkscrews.

Experiment 37: Balancing Corkscrews

For this experiment—in a variation of the well-known variant with floating forks—we find almost all the necessary things on the well-laid table: two corkscrews, the already pulled out cork, the wine bottle ready for pouring and glasses. A wooden toothpick completes the collection of materials.

In the first variant of the experiment, the corkscrews screwed into the cork balance on the edge of the neck of a wine bottle. The balance is so good that this construction is completely unaffected by pouring and continues to balance happily on the bottle neck (Fig. 7.9).

How does this amazingly good balance come about? We need to use the concept of the center of gravity or the center of mass here (see info box). The corkscrews screwed into the cork form a system of three individual bodies with three individual masses. For all three bodies, the individual center of gravity can be determined. But the whole system also has its own center of gravity, which in this case lies outside the three components (Fig. 7.10).

When balancing on the bottle neck, the system's center of gravity is located directly below the support point on the cork. This makes the

Fig. 7.9 The cork with its "weights" continues to balance on the bottle neck even while pouring

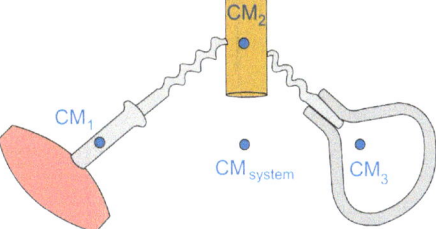

Fig. 7.10 Centers of mass (CM) of the individual parts and the entire system

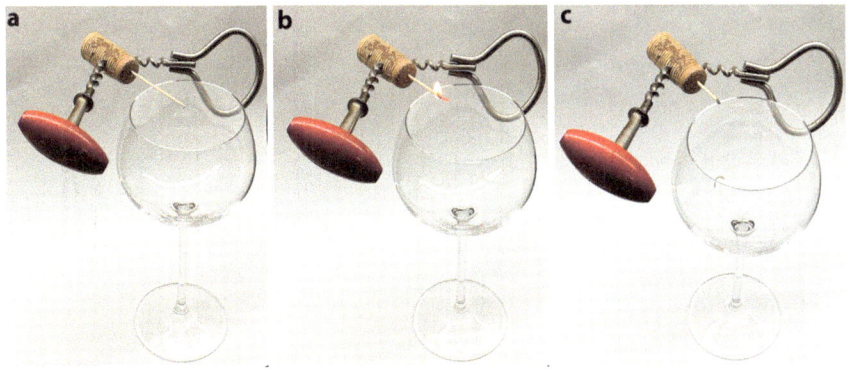

Fig. 7.11 The connected corkscrews continue to balance even after the stick has burned off (a–c)

equilibrium situation stable. We can speak of a *stable equilibrium* when the system is slightly deflected from the resting position and it returns to its original position on its own. We can nudge the pair of corkscrews sitting on the bottle neck a bit. It returns—albeit a bit wobbly—to its position and is therefore stable, even when pouring.

In the second variant of the experiment, we go one step further and the result appears even more spectacular. First, a toothpick is inserted into the cork with the two corkscrews in the direction of the cork axis. The whole thing is then balanced on the edge of a glass using this toothpick (Fig. 7.11a). When the setup then rests on the edge of the glass, this situation alone looks surprising again. But that's not all! If we ignite the tip of the toothpick pointing into the middle of the glass (Fig. 7.11b), it becomes truly astonishing. The stick burns off, but as if by magic, it extinguishes exactly at the point where it rests on the edge of the glass. If the demonstration goes perfectly, then the ash falls off and the corkscrews are still balancing next to the glass—only supported by the very end of the remaining toothpick (Fig. 7.11c).

For the explanation, we again need the center of gravity. Since the tooth-pick was just inserted into the cork as an extension of the axis, the system's center of gravity (Fig. 7.10) now falls approximately in the middle of the stick. If it rests exactly at this point on the edge of the glass, then the system's center of gravity is just below or at the support point and the system is in equilibrium.

The question remains as to how the flame knows to go out at the "right place". This is where the thermal conductivity of the glass comes into play. The burning of the wood requires a certain temperature, which is easily achieved with the ignition. However, as soon as the burning part of the wood reaches the edge of the glass, heat flows into the glass and is conducted away. This "cools down" the flame and it therefore extinguishes exactly at the contact point between the glass and the resting wood.

Center of Gravity

The *center of gravity* of a system is also referred to as the *center of mass*. If the system consists of i components, each with the individual mass m_i, then the location of the center of gravity $\vec{r}_{CM}$ is defined as the sum of the weighted position vectors $\vec{r}_i$:

$$\vec{r}_{CM} = \frac{1}{m_{total}} \Sigma_i m_i \vec{r}_i$$

Here, m_{total} represents the total mass of the system.

For a body with continuously distributed mass (we can imagine a potato here, for example), the individual masses transition into infinitesimal (infinitely small) mass elements dm. The sum from the center of gravity equation then transitions into the integral

$$\vec{r}_{CM} = \frac{1}{m} \int \vec{r} \, dm$$

Static Equilibrium

A body is in *static equilibrium* when the following two conditions are met:

- The sum of the vectors of all external forces acting on the body is zero:

$$\Sigma_i \vec{F} = 0$$

- The sum of the vectors of all external torques acting on the body is zero:

$$\Sigma_i \vec{M} = 0$$

Meeting the first condition is necessary, but not sufficient on its own. It is conceivable that forces could act on a body and cancel each other out in their vector sum. However, if the points of attack are different, a torque acting on the body results from the two forces (see info box of experiment 38). Therefore, only the fulfillment of both conditions is sufficient for a static equilibrium.

Experiment 38: Bottle Holder at the Limit

If you want to draw your guests' attention to the intended wine, a very simple, yet astonishing bottle holder can be recommended (Fig. 7.12). With ease, the bottle seems to hover above the table. Although it is supported by the bottle holder, this stands at a surprising angle on the table. It may also be surprising that the bottle holds at every fill level from full to empty.

This too – as previously with the corkscrews – is a well-balanced act. In a first simple explanatory approach, we consider the bottle and holder as a system, whose common center of gravity must lie exactly above the small contact area of the bottle holder in order for the holder and bottle to stand (Fig. 7.13a). In this case, there is no resulting torque and the system is in stable equilibrium.

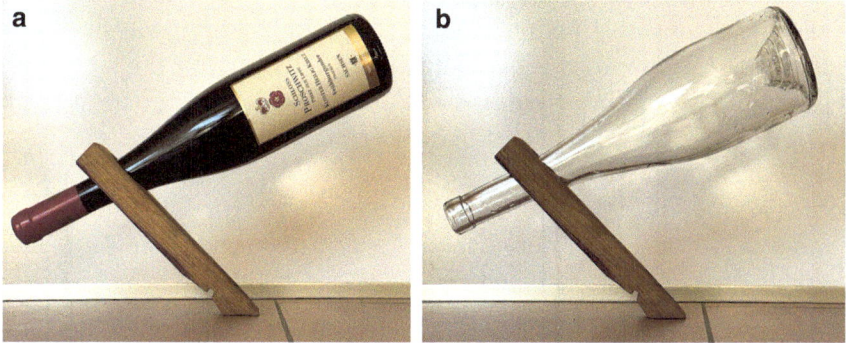

Fig. 7.12 Bottle Holder "in action"; with a full (**a**) and an empty (**b**) bottle

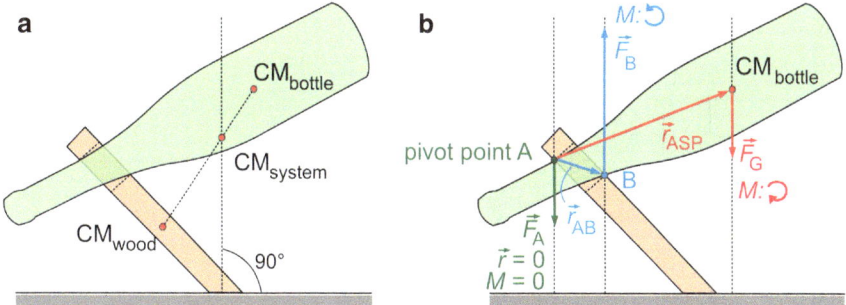

Fig. 7.13 Centers of gravity in the bottle and holder system (**a**), forces, levers and torques on the wine bottle in the holder (**b**); force vectors are not to scale

With this explanation, we could actually leave it at that. Nevertheless, for a slightly more detailed analysis, a consideration of the acting forces and torques (see info box) is interesting from a physical point of view.

Because the bottle and bottle holder are obviously in a stable equilibrium, we can assume that the sum of all torques is exactly zero. But that does not mean that there are no forces and torques present. The bottle and bottle holder are of course each subject to their gravity. If both bodies were independent of each other, the bottle would simply fall down and the bottle holder would tip over from its inclined position. But now the two bodies are mechanically connected, which results in their common stability. Let's take a closer look at the conditions for forces and torques for the wine bottle in Fig. 7.13b:

The storage of the bottle neck in the bore leads to the two contact points A and B, where the forces $\vec{F}_A$ and $\vec{F}_B$ are exerted on the bottle. For further consideration of the torques, we define point A as the pivot point. (We could also define other points for this.) The force $\vec{F}_B$ acting on the bottle at point B has an upward vertical component. Together with the radius vector $\vec{r}_{AB}$ pointing from pivot point A to B, this results in a torque that is oriented counterclockwise. In addition, the weight force $\vec{F}_G$ acts on the bottle. With the radius vector $\vec{r}_{ASP}$ pointing to the center of gravity, another torque is produced, which, however, is oriented clockwise. Both torques necessarily cancel each other out. A force $\vec{F}_A$ also acts on the bottle at the pivot point itself. However, since the radius vector disappears here, no torque results from this force.

With this and the above consideration, it can also be understood why the bottle holder can achieve a stable position with both a full and an empty bottle. Although the center of gravity of the bottle CM_{Bottle} remains approximately in the same position, due to the lower mass, the system center of gravity CM_{System} moves closer to the wood center of gravity CM_{Wood}. To bring the bottle holder back into stable equilibrium, the bottle must be pulled a little out of the holder hole until the system center of gravity is again vertically above the support surface.

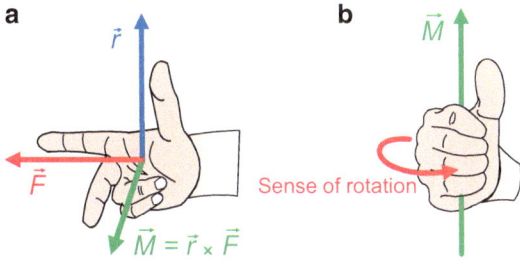

Fig. 7.14   Right-hand rules for the vector product (**a**) and for the orientation of the direction of rotation for the torque (**b**)

Torque

The torque $\vec{M}$ is a vectorial (directed) quantity. It results as the vector product of the radius and the force vector:

$$\vec{M} = \vec{r} \times \vec{F}$$

In the sense of a right-handed system, the torque vector is perpendicular to the radius and the force vector (Fig. 7.14a). The following rule applies to the orientation of the torque: If the thumb of the right hand points in the direction of $\vec{M}$, then the curved fingers indicate the direction of rotation according to Fig. 7.14b (screw- or corkscrew rule).

The magnitude of the torque results for right angles from the product of the magnitudes of radius and force. In the general case (any angle α between radius and force vector) applies:

$$M = r \cdot F \cdot \sin(\alpha)$$

The torque plays a role in many everyday situations. Many might be familiar with the "torque wrench" from changing tires, for example. On playgrounds, the net torque on both sides of a seesaw determines who rises or falls.

From the calculation equation for the torque the unit also results: 1 Nm ("Newton meter").

Experiment 39: The Floating Wine Glass

Is your table too small for all your guests? No problem, here comes a solution to extend the table surface! After we have already made corkscrews float and could present the amazing statics of a bottle holder, we now again defy the laws of gravity—at least Fig. 7.15 gives this impression. As if held by a

Fig. 7.15 A wine glass appears to stand freely and without support on a glass plate

ghost hand, a wine glass filled and pushed far over the edge of the table simply does not want to fall! Its center of gravity is obviously already beyond the edge, over the abyss of the table.

Let's investigate the holding "ghost hand" a bit. The glass plate protrudes with its half over the edge of the table. This alone hardly allows a stable position. But now there is also a decently filled wine glass (mass: approx. 350 g) relatively far out on the "floating" side of the glass plate. So there must be a force that offers a match to the weight force of the wine glass. Since the cause in Fig. 7.15 is hardly visible, the trick must be revealed here: It is the force of the water that makes the glass float. A few drops of water between the table surface and the glass make the glass plate stick firmly enough to the table so that it can withstand the weight on the free side. Can water act like glue? Indeed, one can say that, even though there are better glues than water. In both cases, the adhesive forces between the water or glue molecules and the glass or table plate are sufficiently large. We are talking here about *adhesive forces*. But also the thin film of water between the table and the glass plate must not tear quickly in a glue. This means that cohesive forces must also act between the water molecules. These are the *cohesive forces*.

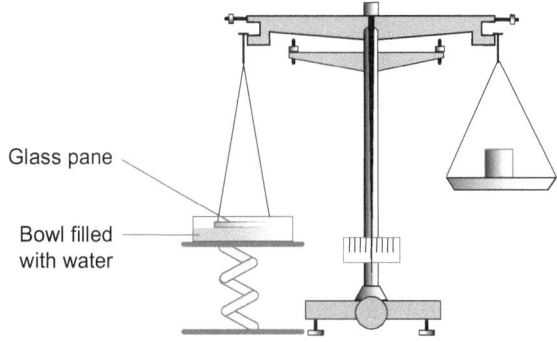

Fig. 7.16   Evidence of adhesive forces between water and glass

Cohesion and Adhesion Using the Example of Glass and Water

Cohesion. Both attractive and repulsive forces exist between the molecules of a substance, which overlap. The ratio of these forces determines the shape and volume of solids or liquids.

The balance of forces particularly determines the distance of the molecules. If one tries to increase or decrease this distance, a corresponding force must be exerted.

Adhesion. Forces also act between the molecules of a solid (e.g., glass) and a liquid (e.g., water) that are in contact with each other. These lead to an attraction of the molecules at the interface. For substances like water, whose molecules have a dipole character, polarity plays a decisive role. In addition, adhesion can also be explained by the displacement of charge carriers in the substances.

Comparison of Adhesion and Cohesion in Water and Glass. The effect of the adhesive forces for the combination of water and glass can be demonstrated in a simple experiment (Fig. 7.16). A small glass plate is attached horizontally to one side of a beam balance. Then the balance is brought into equilibrium. Now a bowl filled with water is brought up to the glass plate from below so that it touches the water surface of the bowl. To separate the glass plate from the water surface, a force is required, which can be determined by placing weights.

If the glass plate tears off from the water surface, it can be seen that the glass plate is still wet with water. This shows that the tearing occurred between the water molecules. The adhesive forces between water and glass are in this case greater than the cohesive forces of the water molecules among themselves.

Experiment 40: Spun, not Stirred …

If you've been too generous when pouring for your guest and want to carry the full glass to them, it's not uncommon for a spill of the fine drop to occur. Admittedly, a good wine is best enjoyed from a stemmed glass, where the risk of spilling due to the small pouring amount is minimal (to allow the full bouquet to unfold as quickly as possible, it is recommended to swirl the glass (see Experiment 9) and therefore only fill it about a third). You are probably already familiar with this problem from carrying filled water glasses, transporting coffee or tea, and even with wine there are certainly situations where a spill is very likely. The "Palatinate tube" has been mentioned several times and in its homeland it is good tradition to fill it even beyond the calibration mark. A spill here is almost guaranteed without tools! This problem also occurs with other wine glasses, e.g. with the Roman glass, which is also filled completely. Before we present a solution to the problem, which also withstands extreme situations (e.g. sprinting, running tight curves or abrupt stopping), we first turn to the cause of the spill. Once we understand the reason for the mishap, the solution to the problem is just a small step away. So what is the cause of the spill when carrying a full glass?

To understand the spill of the wine, we first fill a sealable (!) container about a quarter full with red wine and accelerate it by pushing it horizontally on the table top (Vogt & Kasper, 2022). We follow the movement process with a vertical view or—even better—create a high-speed video with the smartphone. The recording can then be viewed and analyzed at leisure afterwards. In Fig. 7.17, an example of a video snapshot is shown. We can

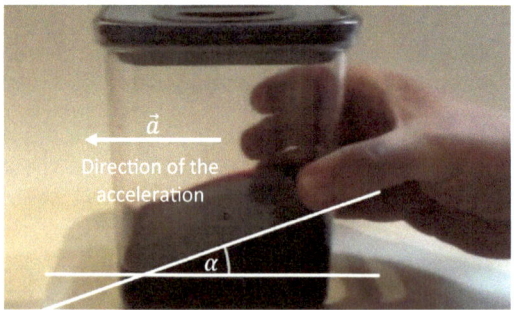

Fig. 7.17 The surface of a liquid accelerated in the direction of translation forms an angle to the horizontal, which depends on the amount of the acceleration

clearly see here that the liquid level—unlike in the state of rest—is no longer horizontal, but is tilted downwards in the direction of acceleration. With an acceleration to the left, the liquid level on the right side of the container increases and would lead to a spill without a seal at a critical acceleration value. If the container was completely filled, even very small accelerations would cause a mishap. But why is the liquid level tilted during acceleration? This is due to inertia, which you are also well familiar with from everyday life and which appears in accelerated reference systems as a so-called apparent force. (Apparent force because, viewed from the outside, there is by no means a force acting on the liquid to the right.) If you accelerate very quickly with your car, your body is pressed into the seat and if you don't hold on tightly enough in the bus, you might even fall backwards when starting. This is exactly what happens to the red wine in the container or possibly also in the overfilled wine glass.

If you want to prevent spilling, you should not fill the glass too full or accelerate it too much when carrying it conventionally. And what if a full glass needs to be transported quickly? After all, we promised you a simple solution at the beginning of the chapter. Then the liquid surface must be kept parallel to the base of the glass in some other way. This can be achieved with a very simple setup, which you must definitely recreate for demonstration purposes at your next party. It's an experiment for the brave, but we can assure you that it will work. At least in our experiments, no glass has ever broken and not a single drop has been spilled! To carry out the experiment, you only need a wooden board (e.g. 18 cm × 25 cm), in whose corners you drill a hole for attaching a string. You bring all four strings, each about 80 cm long, together at one point and tie a knot (Fig. 7.18). You use the suspended board, held in your hand by the strings, as a tray for transporting your glasses. If you now accelerate or decelerate strongly, the "pendulum" swings out and the tray is kept on a circular path by the so-called centripetal force. If this is uniform, then the acceleration points towards the center of the circle and is thus perpendicular to the wine surface. An acceleration in the tangential direction is no longer present and therefore there is no reason for spilling. Even when accelerating or decelerating while carrying, the radial component of the acceleration is significantly greater than the tangential one, as shown by an acceleration measurement via smartphone (Fig. 7.19; Tornaría et al., 2014). So if the butler James in "Dinner for One" had used a tray pendulum, some of the drinks could have been saved when he tripped over the lion's head.

Fig. 7.18 Even with a complete rotation, the liquid surface remains approximately parallel to the base of the glass and nothing spills over

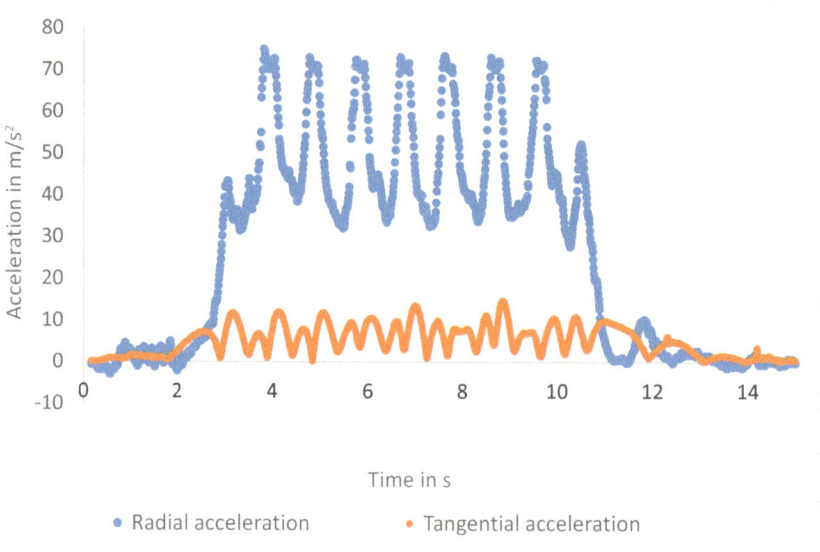

Fig. 7.19 Radial and tangential components of the acceleration measured with the accelerometer of a smartphone instead of the wine glass on the tray during several rollovers

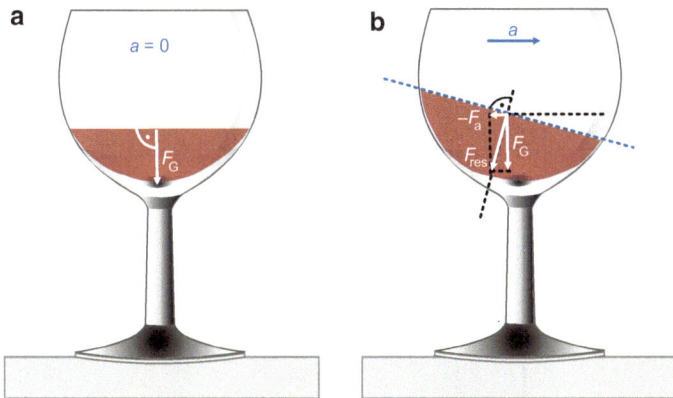

Fig. 7.20 Free surfaces at rest (**a**) and during accelerated translational movement (**b**), based on Sigloch (2014)

Accelerated Translation of Liquids

A liquid generally completely adapts to the vessel walls and forms a so-called *free surface*, (a *liquid level*). Free surfaces are always perpendicular to the resulting force. The simplest example of this is the liquid level of a water glass resting on the table, which forms perpendicular to the gravitational force and approximately—apart from the meniscus at the glass wall (see Experiment 47)—forms a horizontal plane (Fig. 7.20a). In fact, the surface follows the curvature of the earth and is therefore strictly speaking a spherical surface section. However, this aspect can be safely neglected for small free surfaces.

In an accelerated translational movement, the free surface forms an angle α to the horizontal (Fig. 7.20b) and the following applies:

$$\tan(\alpha) = \frac{a}{g}$$

(a: magnitude of the acceleration in the direction of translation, g: acceleration due to gravity)

With this knowledge, we can now also estimate with what acceleration the container filled with red wine was moved to the left. For this, we take the angle a of 19° from Fig. 7.17 and insert it together with the gravitational acceleration of 9.81 m/s^2 into the calculation equation:

$$a = \tan 19° \cdot 9.81 \, \frac{m}{s^2} \approx 3.4 \, \frac{m}{s^2}$$

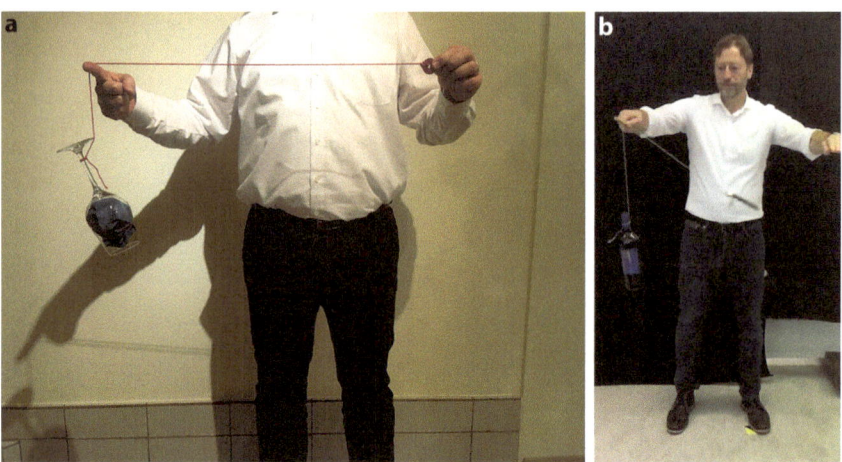

Fig. 7.21 Setup of the described experiment (**a**) and modification with wine bottle, corkscrew and holding rod (**b**)

Experiment 41: The Falling Wine Glass

After spinning filled glasses, we would like to introduce another party trick with the falling wine glass, for which you only need to bring a piece of string to the invitation. The execution of the experiment is very simple, but the result is all the more surprising (Vogt & Kasper, 2021a)! First, one end of the thread is attached to the stem of a red wine glass that is already available, the other to a finger ring or a washer also brought along. With one hand, the smaller body is held and the thread is led over the extended index finger of the other hand. Both hands should be at about the same height, so that the thread initially runs horizontally (Fig. 7.21). Due to the weight force acting on the red wine glass, the thread is taut and the execution of the experiment is limited to releasing the washer. To make the glass clearly visible in the video recording, it was previously filled with a crumpled sheet of blue paper and the washer was colored with a red permanent marker.

First, you can consider what will probably happen in the experiment and only then should you continue reading!

If you assumed that the glass would fall to the ground and a soft pad could be helpful, you are undoubtedly in good company. Especially people who still remember what they learned in physics class are often wrong here, as they argue with the law of conservation of energy roughly as follows: If

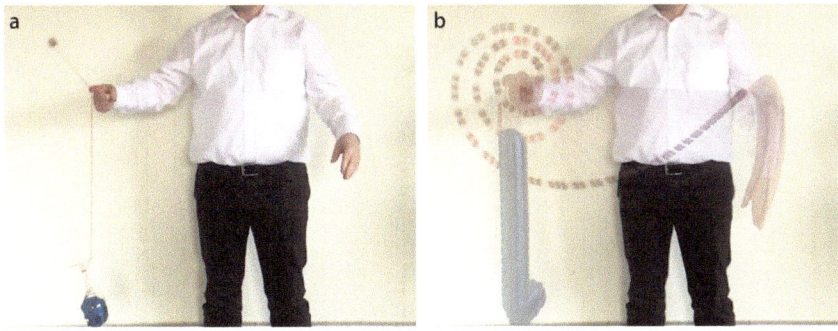

Fig. 7.22 Moment of the first overturn (**a**), stroboscope recording of the entire movement (**b**)

you let go of the washer, the "pendulum" swings back, with the extended finger representing the suspension. The pendulum body (here the washer) reaches its maximum speed in the rest position and on the opposite side—due to friction losses—only approximately reaches its initial height. Without any doubt, due to the law of conservation of energy alone, it is impossible for the washer to overturn, which would be necessary to slow down the falling wine glass!

Contrary to this expectation, however, there is actually an overturning of the released body (Fig. 7.22a) and as can be seen from the stroboscope recording, at least two more (Fig. 7.22b). Due to the multiple wrapping of the finger, the friction force is now so great that the fall of the glass is slowed down and it does not hit the ground. Try it out for yourself!

We do not want to question the validity of the law of conservation of energy at this point. But how can it still lead to multiple overturns and the obviously increasing rotation under its compliance? The reason is provided by the falling wine glass, which leads to a constant shortening of the pendulum length (distance washer—extended finger). As already explained above, the increase in the orbital speed of the pendulum body until reaching the rest position is not surprising. However, there is now an additional increase in speed due to the shortened pendulum length, which can be justified with the law of conservation of angular momentum. If the distance to the center of rotation decreases, the orbital speed must increase to conserve angular momentum, leading to an overturn.

Angular Momentum and Conservation of Angular Momentum

For a point mass, the angular momentum $\vec{L}$ corresponds to the vector product of the position vector $\vec{r}$ and momentum $\vec{p}$

$$\vec{L} = \vec{r} \times \vec{p}$$

and simplifies in the case of circular motion to

$$L = mrv = mr^2\omega$$

(m: mass, r: distance to the center of rotation, v: orbital speed, ω: angular velocity). The orientation of the angular momentum is obtained analogously to the torque using the right-hand rule (Fig. 7.14).

According to the conservation of momentum in translational movements, the law of conservation of angular momentum applies in rotational movements:

$$\vec{L}_{total} = \sum_i \vec{L}_i = \text{constant}$$

($\vec{L}_{total}$: total angular momentum of the system, $\vec{L}_i$: angular momenta of the individual objects). Or in words: The sum of the individual angular momenta is constant in a closed system. For a point mass moving on a circular path, this results in:

$$mr^2\omega = \text{constant}$$

With constant mass, a decreasing distance to the center of rotation is compensated by an increase in angular velocity (cf. increasing orbital speed of the washer in the described experiment).

An interesting application of the law of conservation of angular momentum concerns the Earth-Moon system. Due to tidal friction, the rotation speed of the Earth – although not noticeable to us – is gradually decreasing. This not only leads to an extension of the day duration by about 17 microseconds per year, but also, as a result of the conservation of angular momentum, to a continuous increase in the Earth-Moon distance (approximately 4 cm per year). A phenomenon known from sports that can be understood with the law of conservation of angular momentum is the pirouette effect. This is used in figure skating, gymnastics, and high diving to perform pirouettes, somersaults or twists (Fig. 7.23). At the beginning of the rotational movement, the athlete extends the limbs far from the body. The subsequent pulling towards the body causes an increase in angular velocity.

Fig. 7.23 At the beginning of a pirouette, the figure skater extends both arms and one leg (**a**); after pulling them towards the body, her angular velocity increases significantly (**b**)

Experiment 42: A Break Test for the Fearless!

After the first experiments of this chapter were initially static and then carried out at a relatively low speed, this experiment really depends on speed! We place a 2 m long spruce wood strip on the edge of two wine glasses filled with red wine (Fig. 7.24a), position ourselves in the middle, take a big swing with a stick or a baseball bat and hit the center of the strip with

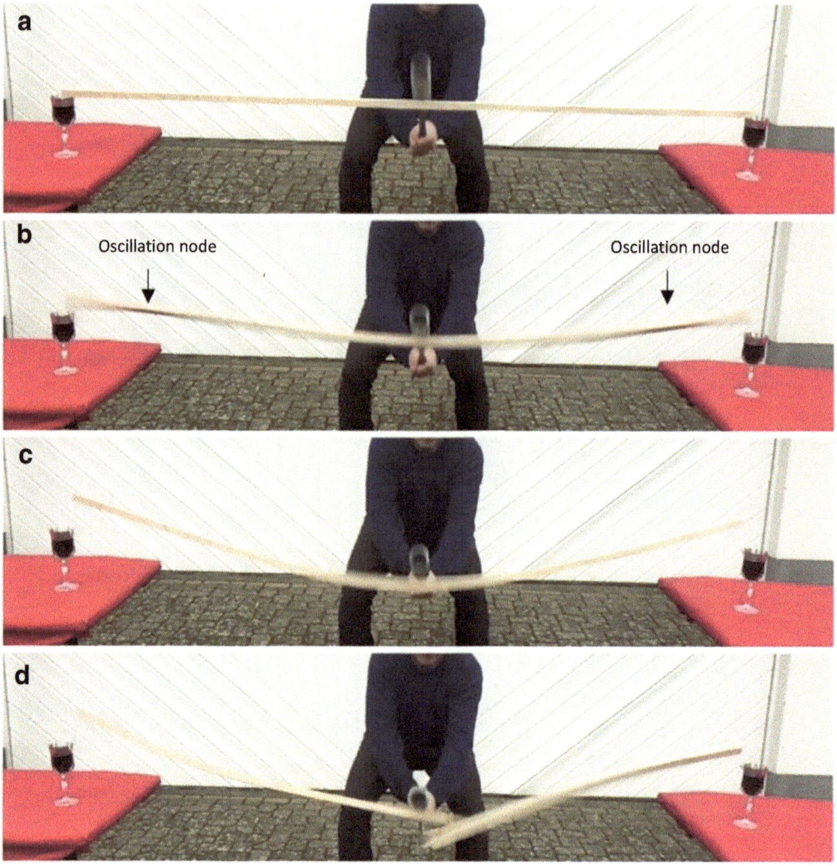

a
b
Oscillation node
Oscillation node
c
d

Fig. 7.24 Consecutive images of a high-speed recording, taken at 240 frames per second

$F \cdot \Delta t$
ω_{CM}
ω_{CM}
v_{CM}
v_{CM}

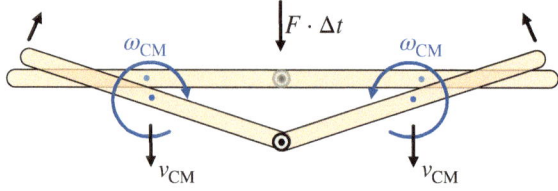

Fig. 7.25 Hinge model of the breaking wooden strip

as high a speed as possible. Intuitively, one would assume that the entire stick will move downwards as a result of the force impulse (see info box for Experiment 2). However, as the individual frames of a high-speed video show, the ends of the stick lift off the glasses without noticeable delay with the deformation of the center (Fig. 7.24b)! Thus, the glasses remain standing and with a bit of luck, not even a drop of wine will be spilled (Fig. 7.24c–d).

To understand this observation, we imagine the wooden slat simplified as consisting of two halves connected by a hinge (Mamola, 1993). The impact causes the hinge to detach and the two halves fall downwards under the influence of gravity. The force impulse leads to a downward initial speed of the centers of gravity (CG) of both halves, but also to a rotational movement around their centers of mass (Fig. 7.25). This rotational movement is the reason for the lifting of the stick ends from the edge of the glass.

In reality, the situation is somewhat more difficult, as the two halves are initially firmly connected and only break due to excessive deformation of the stick. The force impulse on the center of the stick leads to a strong deflection at this point. As with a stone thrown into water, a wave spreads in the wooden strip, which, with a length of 2 m and a propagation speed in spruce wood of around 2900 m/s (see info box), reaches the stick ends after about 0.35 ms. If the ends were still resting on the edge of the glass or not far enough away from it at this time, the vibration of the stick ends would destroy the glasses. To avoid this, the experiment must definitely be carried out with the highest possible impact speed. If you hesitate, you will inevitably need a broom to sweep up the shards!

The experiment certainly requires some practice, so you should rehearse it in a quiet room or even better in your garden before a show experiment. In addition, you can significantly increase your chances of success by taking various measures: 1) First, drive a nail into the stick ends in the longitudinal direction and place the stick with the nails on the edges of the glasses; with a horizontal force, the stick can then slide over the edge of the glass and tipping of the glass is prevented. 2) Place a thin sponge cloth under the glasses (preferably not visible to your audience under the tablecloth), which can at least cushion a small deflection. 3) Use brittle material that breaks as easily as possible. For example, you can saw a strip from a chipboard. If you take these points into account, nothing can really go wrong with the experiment. We did not consider these recommendations, which resulted in one or two glasses breaking during several repetitions of the experiment (Fig. 7.26). Due to flying glass and wood splinters, you should definitely wear safety glasses for this experiment.

Finally, we would like to point out an interesting effect that can also be observed in this experiment: The spreading impulse is reflected at the stick ends and there is an overlay of a forward and backward wave. Analogous

Fig. 7.26 The wooden strip initially lifts off the edge of the glass (top right), but then accelerates downwards overall due to the lack of breakage in the center

to oscillating air columns (see info box Experiment 1), standing waves also form in oscillating rods. You can clearly see these in Fig. 7.24b, where maximum deflections are present in the center and at the stick ends (vibration bellies). In addition, you see two areas where the stick remains approximately at rest (vibration nodes).

Speed of Sound in Solids

Sound can propagate in solids as a longitudinal- or as a transverse wave. In the longitudinal case, the particles oscillate parallel to the direction of propagation, in the transverse case perpendicular to it. The bending vibration excited by the impact leads to a transverse wave, whose propagation speed can be calculated as follows (Lüders & von Oppen, 2008):

$$c_{tr} = \sqrt{\frac{E}{\rho} \frac{1}{2(1+\mu)}}$$

The modulus of elasticity E and the Poisson's ratio μ are parameters from strength theory and are for spruce wood at $E = 11 \cdot 10^9$ N/m^2 or $\mu = 0.4$. With a bulk density of 470 kg/m^3, after inserting the numerical values, the transverse speed of sound is approximately 2900 m/s.

8

Finished Drinking? Experimenting with Residual Alcohol

The last eight experiments are our offer for a "physically supervised clean-up". You will hardly read that it is time for this on the "wine clock", but it does lead us to other insights. When bottles and glasses are emptied, it's time for final truths. These also include the answer to the question of how fit one is to drive. We do not leave the determination of the blood alcohol level exclusively to the police. We measure it ourselves! Did corks get into the bottle during the experimental wine tasting? We'll get them out again. Promise! And if you still feel like "action" after all the experiments, we are of course delighted and recommend clearing up in an unconventional order. And at the very end, all that remains for us is to extinguish the last candlelight—of course not without physical ulterior motives!

An Experiment with the Last Sip

Experiment 43: The Paradoxical Wine Clock

Just before all bottles are completely emptied, you should definitely reserve the last sip of red wine of the evening for another experiment. In addition to this last remainder, you only need two identical shot glasses (which you may find on your table at a later hour anyway), some tap water of approximately room temperature, and a piece of foil. First, fill the two shot glasses to the brim with the red wine and tap water respectively (Fig. 8.1) and pose the following task to your guests: Without using a third vessel, the two liquids

Fig. 8.1 Preparation of the experiment; one of the two shot glasses is completely filled with red wine (**a**), the other with water (**b**)

should change glasses! If the experiment is not yet known, no adequate solution will certainly be presented by the attendees, and you can undoubtedly amaze your table company with the experiment.

First, seal the glass filled with water by placing the piece of foil on it. As we could already see in Experiment 28, the glass now also holds tight when turned upside down (Fig. 5.3) and you can then place this with the opening facing down on the red wine glass (Fig. 8.2a). Now gently pull the foil to the side so that a small opening is created between the glasses at one point. Immediately, a flow sets in that transports the wine into the upper glass (Fig. 8.2b). But the upper glass is completely filled with water, i.e., in return, water must get into the lower glass. This actually happens and can already be clearly observed in Fig. 8.2c; the water flowing into the red wine glass settles at the bottom of the glass. Now it just takes a little time until most of the liquids have changed their position and thus the glass. In Fig. 8.2d, there is still some red wine in the lower glass, but the upper one is already consistently deep red. The experiment is reminiscent of an hourglass, although our "wine clock" may seem paradoxical. How is it possible that the red wine moves against the force of gravity?

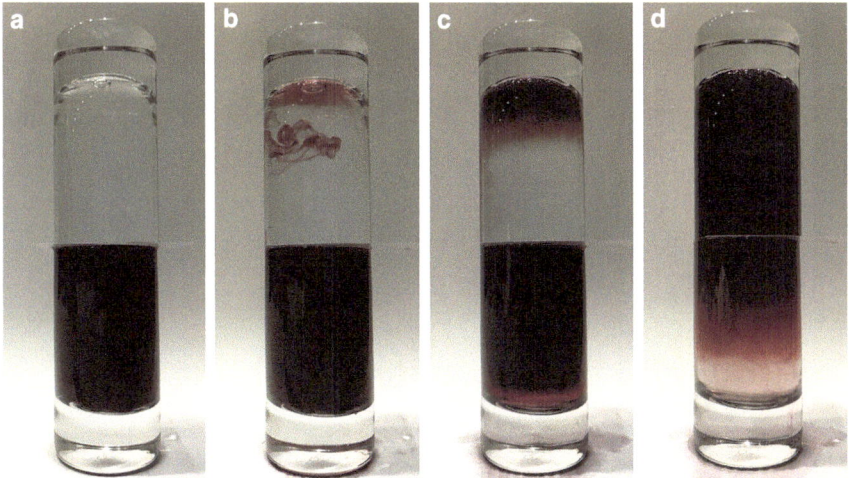

Fig. 8.2 The lighter red wine rises through the small opening and swaps places with the water

The density of alcohol is significantly lower at 789 kg/m³ than that of water at 998 kg/m³. These values refer to a temperature of 20 °C and result in a weighted average of 972 kg/m³ for a red wine with 13% Vol. Due to other ingredients, the actual density is higher, but still less than that of water. This difference in density leads to the onset of free convection (see info box) and thus to the exchange of liquids. At the glass walls, the flow velocity is significantly lower due to adhesion (see info box 39), which can be seen in the middle part of the red wine glass by the low color intensity; there is still some red wine at the edge, whereas in the middle of the glass there must already be predominantly water. This is the only way to explain the already complete coloring of the upper glass (Fig. 8.2d).

Free and forced convection

When a fluid (gas or liquid) with different densities is in a gravitational field (e.g., in the Earth's gravitational field), a flow sets in as a result of the acting buoyancy force, which transports the fluid with lower density upwards. If no other forces act in addition to the buoyancy force, this is referred to as *natural* or *free convection*. The different densities are usually due to temperature differences or different substances. The latter case, which also applies to the experiment described, is called *chemical convection*. Examples of free convection in everyday life include thermals (updrafts due to stronger heating of near-ground air layers) or the Gulf Stream. This transports warm surface water from the Caribbean to the northern European coasts. Its main driving force is—in addition to an increasing salt content—polar ice masses, under which cold and thus relatively dense water sinks from the surface to the seabed. This results in a circulation that extends to the Gulf of Mexico (Fig. 8.3).

If the flow is caused by an external force (e.g., by a pump), this is referred to as *forced convection*. Both forms occur together in part, including in the heating of a residential building. The heating of the heating water alone initiates a natural convection, which could supply the radiators of the entire house. However, the flow is additionally amplified with a pump. The ratio between free and forced convection can be described with the *Archimedes number*.

Blow the Whistle! Acoustic Alcohol Test

Experiment 44: Why Drunk People Whistle Higher

When the party has come to an end and you want to make your way home, the question arises who should get behind the wheel and who might still be able to take a seat in the back of the car. To put it very clearly: Of course, you should definitely clarify this question at the beginning of the evening and the chauffeur should then completely abstain from consuming alcohol! The following experiment works all the better then.

For this, you only need a whistle, e.g., a shrill whistle (trill whistle without a ball), a boatswain's or a dog whistle (Fig. 8.4). First, a sober person with a blood alcohol concentration (BAC) of 0.0 g/l blows into the whistle and you determine the frequency of the fundamental tone with a suitable

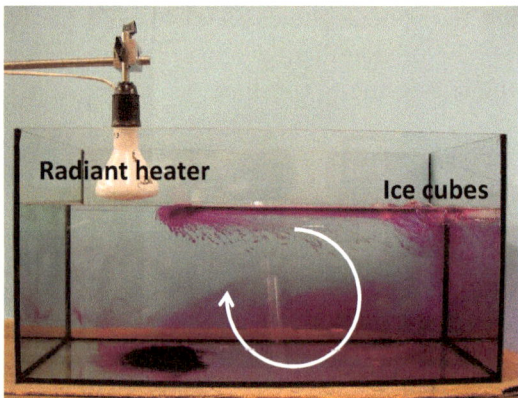

Fig. 8.3 Analogous experiment to the Gulf Stream; the water cooled by the ice cubes sinks, warm surface water flows from the left. A circulation occurs, which was made visible in the experiment by potassium permanganate

Fig. 8.4 Dog whistle used for the experiment

app. How this works and which apps are suitable for this, we have already presented elsewhere, e.g., in experiments 1 and 15. The determined fundamental frequency is noted and compared with the pitch that other party guests produce. You will find that the frequency of the whistle tone increases with the blood alcohol concentration. But why do intoxicated or drunk people whistle higher than sober people?

The propagation speed c of waves is obtained by multiplying the wavelength λ by the frequency f (see info box). The wavelength of the fundamental tone of a whistle depends solely on its geometry and is, for example, four times the length of a tube closed on one side (see Experiment 18) and twice the length of a tube open on both sides. The wavelength for the used dog whistle cannot be estimated that easily, but we can determine it by simply using the measured fundamental frequency in the sober state. The following applies:

$$\lambda = \frac{c}{f} = \frac{351 \text{ m/s}}{4094 \text{ Hz}} \approx 8.6 \text{ cm}$$

Here we have taken into account the speed of sound in air at 36 °C (body temperature), which is not entirely correct. Due to breathing, 4% Vol. of the 21% Vol. oxygen present in air reacts to CO_2, which, however, only has a very small influence on the speed of sound and can therefore be safely neglected. The wavelength of the fundamental tone of the used whistle is therefore 8.6 cm and is constant due to the clear dependence on the whistle geometry. A change in the whistle frequency can therefore only be caused by a changed speed of sound of the gas flowing through the whistle. Rearranging the equation for the frequency results in:

$$f = \frac{c}{8.6\,\text{cm}}$$

From the increase in whistle frequency with alcohol consumption, we can therefore conclude that the speed of sound of the exhaled air increases with the blood alcohol concentration. This can only be explained by the fact that the speed of sound of ethanol vapor—its proportion in the exhaled air increases with the BAC—is greater than the speed of sound of air.

To examine the dependence of the whistle frequency on the BAC more closely, we conducted the following experiment: A test person drank six half-liters (0.5 l beer) at a steady pace over a period of 3.5 h, solely for scientific purposes. After each beer, the frequency of the fundamental tone was determined three times and the average was calculated. In addition, the BAC was measured each time with a commercially available digital breathalyzer. The result of the measurement series is shown in Fig. 8.5. As already explained, the frequency increases with the blood alcohol concentration, with the linear regression providing a coefficient of determination of 0.75 and the following dependence:

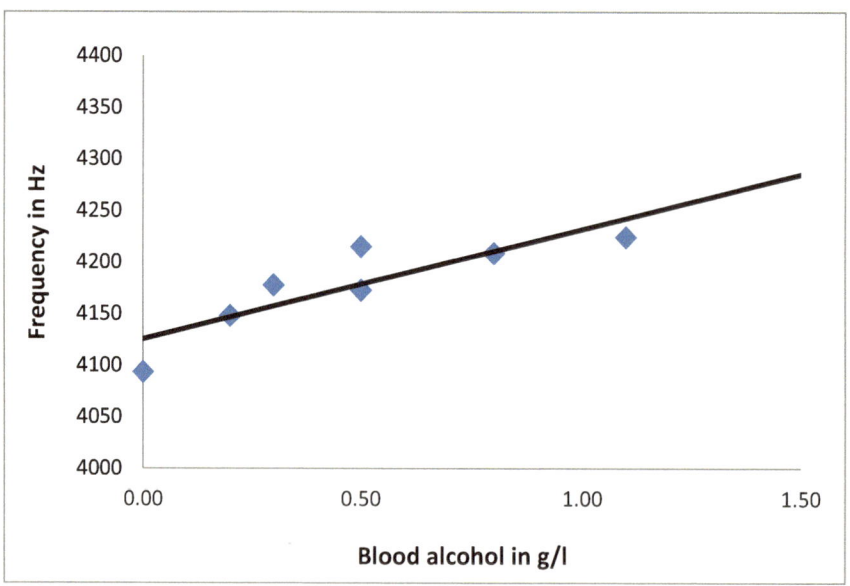

Fig. 8.5 Fundamental frequency of the whistle in relation to the blood alcohol concentration

$$f = \underbrace{106\,\text{Hz}}_{\approx 100\,\text{Hz}} \cdot BAC\frac{1}{\text{g}} + \underbrace{4125\,\text{Hz}}_{\approx f_0}$$

This relationship leads to a surprisingly simple rule of thumb:

$$BAC \approx \frac{\Delta f}{100\,\text{Hz}}\frac{\text{g}}{1}$$

The frequency change in Hz divided by 100 provides the blood alcohol concentration in g/l. However, the equation could not be exactly reproduced in repeat experiments, so you will still have to "blow" and not "whistle" in traffic checks in the future!

Nevertheless, the experimental approach could be usefully applied in the context of "wine". In European wine-growing regions, fermentation gas accidents are becoming increasingly rare due to exhaust systems and corresponding warning systems. However, in less advanced regions, it would be possible to identify a critical CO_2 content in the air via the whistle frequency. For this purpose, an app could be programmed in which the whistle frequency at a certain temperature and normal air composition would be stored. After whistling in the wine cellar, the app could then display whether the CO_2 content in the breathing air exceeds a critical value based on the changed natural frequency and taking into account the temperature read out via the internal smartphone sensor.

Origin and Propagation of Sound Waves

Three prerequisites are necessary for the propagation of sound waves:

1. A multitude of oscillators capable of vibration of the same natural frequency is required.
2. The oscillators must be coupled, i.e., forces must be able to act between them.
3. At least one oscillator must be stimulated to vibrate.

You can perform a simple model experiment with clothespins that you hang on a gift ribbon (Fig. 8.6). All clothespins are identical and therefore have the same natural frequency f or period T. If you now deflect one of the clothespins, this energy is transferred to the next and a wave propagates. The propagation of sound waves in gases or in a solid body is completely analogous—the individual oscillators are then the molecules of the body under consideration. The "clothespin model" can also illustrate basic terms and calculation equations: The distance between two clothespins in the same state of vibration (same

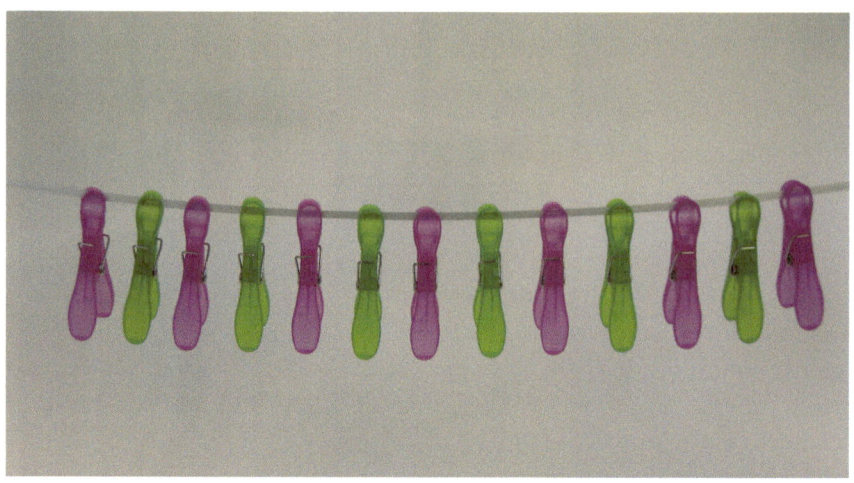

Fig. 8.6 Deflection of one of the hung clothespins leads to the creation of a wave

deflection and same direction of movement) corresponds to the wavelength λ. To cover this distance, the wave needs exactly one period T ($T = \frac{1}{f}$), i.e., the propagation speed c is given by:

$$c = \frac{\lambda}{T} = \lambda \cdot f$$

This relationship applies not only to the clothespins considered, but to any type of wave propagation and especially also for the sound waves in the whistle used in the experiment.

How does the Cork Get Out of the Empty Bottle?

Experiment 45: Cork Release through Friction

Perhaps you have opened a wine bottle unconventionally, so that the cork has ended up inside the bottle. Perhaps the "proper" uncorking has failed and half the cork has slipped into the bottle's belly. This could be easily remedied by decanting. But still, you want to get the cork out again because you like the bottle or because you still need it for experimenting? No problem! This is a relatively easily solvable task.

All that is needed for this is a small plastic bag or even a durable cloth handkerchief. How should you proceed?

The plastic bag can be rolled up and inserted into the bottle up to the opening. If necessary, you can help with a cooking spoon handle. Then, by shaking the bottle a little back and forth, the cork is brought into a position near the bottle neck and so that it is already in the correct longitudinal alignment between the glass wall and the bag. To better fix the cork in this position, some air is blown into the bag (Fig. 8.7b). In the case of using a handkerchief, tie a knot in the tip inserted into the bottle. Now you can pull on the bag. Initially a bit careful, until you can "grab" the cork at the entrance of the bottle neck with the bag. Then pull hard and … the cork is out!

How can this succeed so easily? Doesn't the attempt seem hopeless, as in this way, in addition to the already very tightly fitting cork, additional bag material must also be squeezed through the bottle neck? All the more astonishing is the very high success rate of this method. The physical explanation lies in the phenomenon of friction. More precisely, we are dealing here with both static friction and sliding friction (see info box).

Friction is—and this can be said with good conscience—vital. If it did not exist, and this is just one example among many, we humans would not get anywhere in everyday life; neither on foot nor with vehicles. Shoe soles would glide without resistance and vehicle tires would "spin". Of course, corks would also not hold in wine bottles! The fact that they sit so firmly in the bottle neck and thus protect the wine from spoiling is initially due to the *static friction* between the cork material and the glass of the bottle. Once the

Fig. 8.7 A bag almost effortlessly transports the cork out of the empty bottle

cork starts to move when pulling the cork, the static friction is indeed overcome. But then the *sliding friction* still opposes the all too easy pulling out.

The friction between two bodies depends on the one hand on the force with which the two friction partners act on each other. In everyday life, it is often the weight force that is the cause—e.g. the runner of a skate exerts a force on the ice that corresponds to the weight force of the person standing above it.

In addition, both static and sliding friction depend on so-called friction coefficients. These are material-specific. And this is exactly where the trick with the bag comes into play: If the cork and bag are in the correct position in the bottle, then part of the cork is surrounded by the plastic bag, another part of the surface of the cork is in close contact with the glass of the bottle neck. When the bag is pulled, three different static friction forces oppose the attempt to pull it out. There is friction at the respective contact points between cork and plastic as well as between plastic and glass and between cork and glass. The normal forces acting in all cases are the same, but the static friction coefficients differ. The friction to be overcome between plastic and cork material is clearly the greatest. The cork thus "sticks" to the bag and is pulled out with it. The conditions are quite analogous for the sliding friction.

Adhesive and sliding Friction

The frictional force between two bodies is proportional to the normal force F_n, with which a body acts on its base. The proportionality factor is called *friction coefficient* μ. Thus, the frictional force F_f can be specified:

$$F_f = \mu \cdot F_n$$

The friction coefficient is dimensionless and depends on the combination of material types as well as their condition (e.g., roughness).

In the static case, the *static friction force* F_{sf} opposes the force trying to set the body in motion. If the body starts moving on its base, then the *sliding friction force* F_{slf} opposes this movement. For both types of friction, the friction coefficients differ:

$$F_{sf} = \mu_{sf} \cdot F_n \text{ and } F_{slf} = \mu_{slf} \cdot F_n$$

For the same material combination, the coefficient of sliding friction is always smaller than the coefficient of static friction. For sliding friction, it is almost independent of the speed of the body sliding on the base. Sliding friction is always associated with a thermal conversion of kinetic energy.

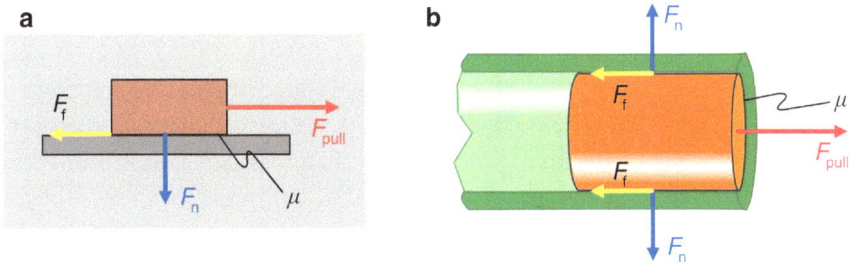

Fig. 8.8 Friction force between two bodies in general (**a**) and in the case of a bottle cork (**b**)

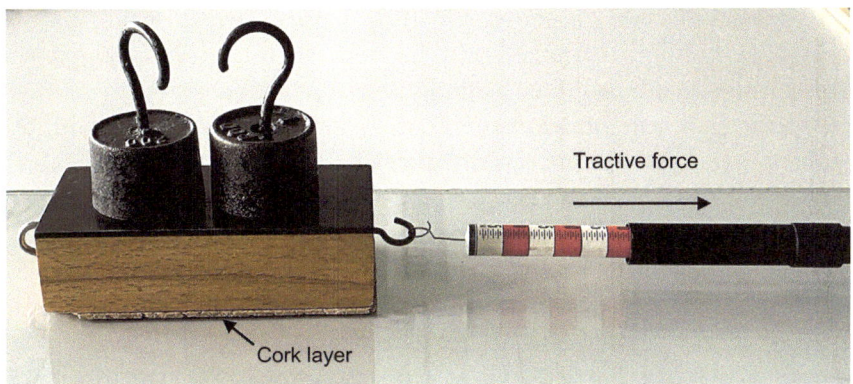

Fig. 8.9 Determination of the friction coefficients for cork-glass in a simple experiment

So, how large is the friction coefficient between cork and glass? Figure 8.8a already gives us the crucial experimental hint: A pulling force F_{pull} is exerted on a body with a cork coating on the underside and the known mass m. The pulling force, which is just enough to set the body in motion, thus also just overcomes the static friction force F_{sf}. This force can be determined with a spring force meter (Fig. 8.9). The normal force F_n results from the measured mass: $F_n = m \cdot g$. With the equation for the friction force (see info box), the friction coefficient can then be determined.

$$\mu_{sf} = \frac{F_{sf}}{F_n}$$

The determination of the sliding friction coefficient proceeds analogously. The pulling force required to maintain a constant sliding speed is F_{slf}.

From our measurements, the following values result for the cork-glass combination: static friction coefficient $\mu_{sf} = 0.70$; sliding friction coefficient $\mu_{slf} = 0.63$. These values are relatively large compared to friction coefficients for other material combinations and are on the order of the friction numbers of rubber tires on asphalt. In both cases, large friction numbers are desired. After all, the cork should sit tight! For comparison, a case where the friction should be as small as possible: The steel runner of a skate has a static friction coefficient of about 0.03 and a sliding friction coefficient of about 0.01.

The friction with glass is not the only advantage of this natural material. Even if cork is much less in use today—what makes it so unique?

Its special physical properties are a low density—which makes cork very light as a building material—and its poor thermal conductivity, which makes it so attractive as an insulation material or for floor coverings.

Specifically for its function as a bottle stopper, cork can shine with other properties (Fig. 8.10): The cell structure of cork with its elastic membranes allows it to flexibly respond to pressure and restores its original shape after the force is applied. Cork contains the natural polymer *suberin* (*Quercus suber* is the cork oak) stored in its cells. Cork thus also provides a natural and dense barrier for moisture and liquids. This applies to a limited extent even for gases. Limited because some gas molecules in the air manage to overcome the cork barrier. This is the reason why wines can continue to mature and of course spoil in bottles sealed with cork due to very small, but

Fig. 8.10 Natural material "Cork"

steady oxygen supply over months or even years. In addition, cork is anti-microbial, which is why you have probably never discovered mold on bottle corks.

And finally, we return to friction. Cork is considered a resistant "friction partner". This means it wears less and slower than other materials. This actually speaks for using it multiple times. In chemical laboratories or pharmacies, cork stoppers were indeed a standard material for a long time before the advent of silicone stoppers.

In the next experiment, we will tackle the same problem of the cork trapped in the bottle again—with a completely different approach.

Experiment 46: Cork Release by Buoyancy

Admittedly, this variant is a bit more complex, but from a physical point of view no less interesting and also suitable as a party trick. To remove the cork from the already emptied bottle (Fig. 8.11a), it must first be filled with water up to the opening. Now we seal the bottle with a finger and tilt it so that the cork floats in an upright position up to the bottle neck. After placing the bottle upright, the actual trick comes: A second cork is pressed into the bottle neck with a lot of force (Fig. 8.11b), but only so far that it can still be easily pulled by hand without a corkscrew. The cork then floats as if by magic a bit into the bottle neck. The second cork is pulled, the bottle is refilled with water and the process is repeated until the lower cork is firmly

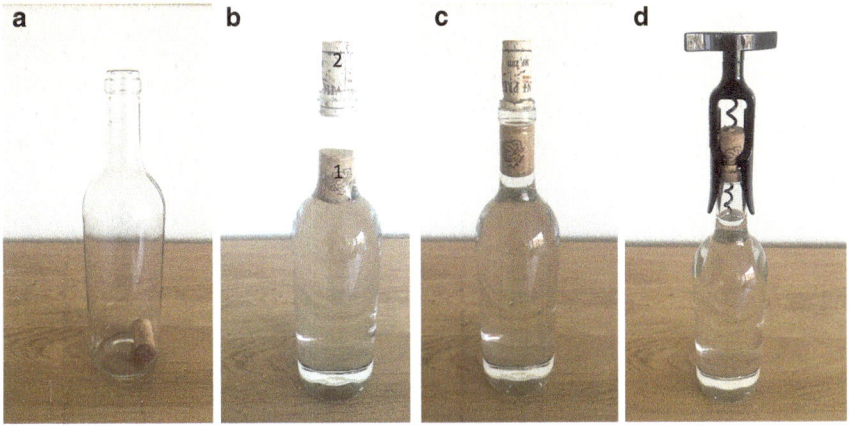

Fig. 8.11 The cork is removed from the bottle belly by utilizing the buoyant force (a–d)

in the bottle neck (Fig. 8.11c). Now the cork can be removed in the conventional way with a corkscrew and the bottle can be emptied (Fig. 8.11d).

But why does the cork move into the narrow bottle neck during the pressing? The cause of this is the incompressibility of the water and the lower density of cork. As a result of the force exerted with cork 2, the pressure in the liquid increases. The water itself cannot be compressed, but cork 1 can due to air inclusions. If we press cork 2 with the volume ΔV into the bottle neck, the volume of cork 1 must necessarily decrease by exactly this amount. If its diameter becomes smaller than that of the bottle neck, it rises due to the buoyant force acting on it. If we now reduce the pressure in the liquid by pulling out cork 2, cork 1 expands again and in turn presses against the bottle neck. It remains in the position taken.

Archimedes' principle

Any body that is fully or partially in a fluid (gas or liquid) with the density ρ_l experiences an upward directed *buoyant force* $F_{buoyant}$. This corresponds to the weight force of the fluid volume V_{dis} displaced by the body (Archimedes' principle) and thus amounts to

$$F_{buoyant} = \rho_l \cdot V_{dis} \cdot g$$

(*g*: acceleration due to gravity). The cause of the buoyant force is the increasing gravitational pressure with depth of the fluid (see info box of Experiment 36). Forces due to the existing pressure act on all sides of the submerged body, which we simplify as a cube (Fig. 8.12). The lateral forces F_a and F_b are of equal magnitude and compensate each other. However, the forces F_c and F_d, which act on the top and bottom of the body, are different: Due to the greater pressure at the bottom, the upward force is greater, so the resultant also points upwards.

The buoyant force $F_{buoyant}$ corresponds to the difference $F_d - F_c$. With the definition of pressure $p = \frac{F}{A}$ (see info box of Experiment 3) follows

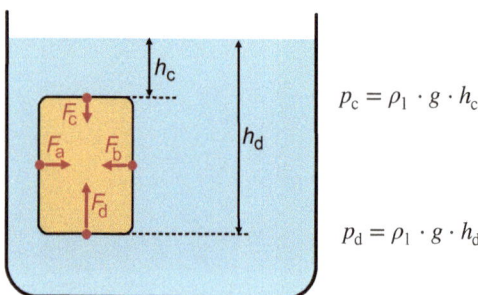

$$p_c = \rho_l \cdot g \cdot h_c$$

$$p_d = \rho_l \cdot g \cdot h_d$$

Fig. 8.12 Derivation of the buoyant force

$$F_{buoyant} = A \cdot (p_d - p_c) = A \cdot \rho_l \cdot g \cdot (h_d - h_c)$$

and, since $A \cdot (h_d - h_c)$ corresponds exactly to the displaced volume,

$$F_{buoyant} = \rho_l \cdot V_{dis} \cdot g = m_{dis} \cdot g.$$

By the way, it is a widespread misconception that a sinking object, e.g., a stone at the bottom of a lake, does not experience a buoyant force. Whether a body sinks, floats, or swims depends on whether the buoyant or gravitational force acting on it predominates. The following conditions apply:

Sinking	$F_G > F_{buoyant}$	$\rho_l < \rho_b$
Floating	$F_G = F_{buoyant}$	$\rho_l = \rho_b$
Swimming	$F_G < F_{buoyant}$	$\rho_l > \rho_b$

Does the Cork not Know Gravity?

Experiment 47: Paradoxical Cork in a Glass of Water

The cork salvaged from the empty wine bottle in the last experiment can now be helpful for the observation of a seemingly paradoxical behavior. If we fill a glass not to the brim with water and put the cork in, we observe that it always strives towards the edge. The cork repeats this behavior even when we try to push it from the edge to a middle position. Where does this striving for the edge come from?

The "physical view" sharpened during our wine tasting helps here. To show the effect a little more clearly, a slice can be cut off the cork, which then floats flat on the water. A close observation reveals that the cork slice not only strives towards the glass edge, but also moves "uphill" in the process (Fig. 8.13a). The water surface visibly clings to the inner edge of the glass, thus forming a slope. The *hydrophilic,* or "water-loving" property of the glass is still in bad memory in connection with the spilling of the tablecloth when pouring (see Experiment 27). In the glass, this property causes a concave curvature of the water surface. But this means that the cork climbs uphill on its way to the glass edge. Doesn't it know anything about gravity?

Oh yes, it knows it quite well. The much greater gravity that acts on the water surrounding the cork is what leads to the buoyancy of the cork and explains its seemingly paradoxical behavior. It is the water with its much greater density (factor 6.7) that pushes the cork upwards (see Info box "Archimedes' Principle", Experiment 46).

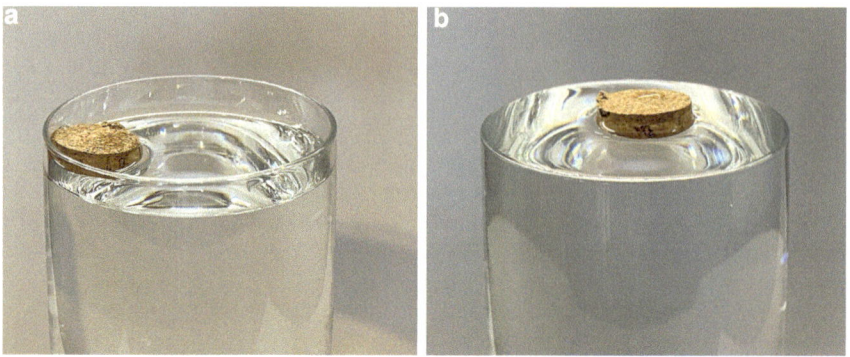

Fig. 8.13 Floating cork slice; in the not full (**a**) and in the "overflowing" glass (**b**)

What if the water surface were convex, i.e., curved upwards? Then the cork should strive upwards and thus towards the center of the glass. We can actually check this assumption relatively easily by filling the glass to the brim and then carefully continuing to fill it with water until the surface actually curves upwards. The cork slice, which has so far been "sticking" to the glass edge, detaches as soon as the upward curvature occurs and moves—again swimming uphill—to an approximately central position (Fig. 8.13b).

Surface tension and meniscus of water in glass vessels

A glass not filled to the brim with water shows a concave curved surface at the edge. Water "wets" glass (see info box for Experiment 27). This means that the attractive *adhesion forces* (between glass and water molecules) are greater than the attractive *cohesion forces* between the water molecules themselves. The water rises slightly at the point of contact with the glass. The surface forms a *meniscus*. This meniscus must be taken into account when reading scales, for example on measuring vessels.

If the glass is filled to the brim and further filled beyond that, it is possible to create a convex surface (a "water mountain"). This phenomenon can be explained with the surface tension of water (see info box for Experiment 28). The attractive cohesion forces between the water molecules act parallel or tangential to the interface between water and air. As a model, one can imagine the water surface as a "membrane" that withstands even small loads (the cork in the experiment or water striders in nature).

By the way, the bulging surface in the cork experiment works better with water than with wine or even high-proof drinks. For comparison, the values of the surface tension σ (values for 20 °C): Water: 0.073 N/m; Ethanol: 0.023 N/m.

Who Pays the Bill? An Experiment à la Otto von Guericke …

At the end of even the most pleasant and wine-filled gathering, the question sometimes arises as to who will pick up the tab. Here, a study of the innovative and versatile *Otto von Guericke* (1602–1686) can help in two ways: on the one hand, of course, because of his qualities as a physicist and science entertainer, but on the other hand also because of his skills as Mayor and City Treasurer of Magdeburg. Even though the main drive of his scientific pursuits were rather philosophical considerations, he occasionally managed to connect science and financial support from third parties. Thus, after a convincing staging of his novel air pressure and vacuum experiments before the Emperor and Princes at the Imperial Diet in Regensburg in 1654, he was able to sell some of his experimental equipment to the enthusiastic Elector of Mainz. This probably included not only the equipment, but also—perhaps even primarily—Guericke's experimental ideas.

And these ideas were indeed significant. They were about nothing less than explaining all phenomena attributed to the *horror vacui* (nature's aversion to emptiness) through air pressure. The most important achievement of *Guericke* in this regard was initially the construction of his air pump, with which he was able to pump out the air from metal vessels after an initially unsuccessful attempt with a wooden barrel.

Subsequently, one of the experimental ideas became world-famous under the name "Magdeburg Hemispheres".*Otto von Guericke* demonstrated them several times, always very effectively for the audience. Using a pump he also developed, he removed the air from two metal hollow sphere halves placed together with a sealing ring. These had the diameter of a *Magdeburg Elle,* which is approximately 42 cm (later, von Guericke also conducted further experiments with spheres of larger diameters up to 60 cm). After the air was pumped out, the two halves could only be torn apart—if at all—with the force of twice eight horses pulling against each other, which then happened with a loud bang. This impressed the crowd at the time, and reconstructed experimental procedures are still spectacular and captivate the audience today.

A small note should be made here about the Magdeburg sphere experiment. *Otto von Guericke* had up to 16 horses working in his demonstrations (Fig. 8.14). That certainly makes an impression! He probably knew it: Half would have done it from a physical point of view! If one half of the sphere

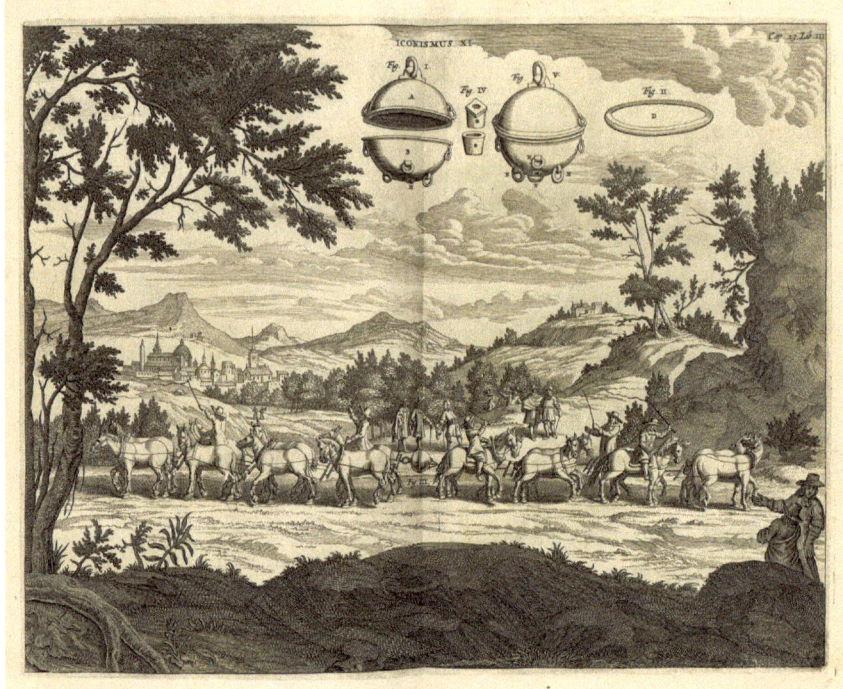

Fig. 8.14 Otto von Guericke: Experimenta nova (ut vocantur) Magdeburgica de vacuo spatio. (Engraving by Caspar Schott, 1672)

had been fixed to a stable point of a building or a mighty tree, eight horses could have been saved, which took on the role of a "counter bearing". But it would have been only half as spectacular!

Experiment 48: New "Magdeburg Wine Glasses"

With a little modesty regarding the "large equipment", we can also perform this experiment at the table. We need two of the emptied wine glasses for this. It is important that they are identical wine glasses and that their rims fit exactly on top of each other. We also make the necessary sealing ring from chamois leather or thick blotting paper by cutting a circular piece with a hole from it. We certainly don't have a pump at hand, and it wouldn't make much sense anyway, as the wine glasses lack any suction nozzles. We create the pressure difference between the inside of the "wine glass sphere" and the air pressure differently: The prepared sealing ring is moistened and placed on the rim of one of the two glasses. A small piece of crumpled paper is ignited in this glass. A small cotton ball is also very suitable. Shortly afterwards, the second glass is inverted and placed as exactly as possible rim to rim over the first glass (Fig. 8.15).

The air that cools down after the flame goes out contracts, creating a vacuum compared to the air outside the glasses. As a result of this pressure difference, the external air pressure presses the two glasses together. Both halves of the sphere can be lifted by the handle of the upper glass without separating. Thus, we can confidently refer to them as "Magdeburg wine glasses". The principle is exactly the same: air has mass—1 l of air weighs about 1 g—and at sea level, it exerts a standard pressure of 1013 hPa on us. This corresponds to about 10 N per cm².

If the experiment succeeds in creating such a pressure difference that the force related to the opening area exceeds the weight force of the hanging glass, then the two glasses can be lifted together adhering to the upper glass. However, how many horses are needed for separation still requires empirical verification ...

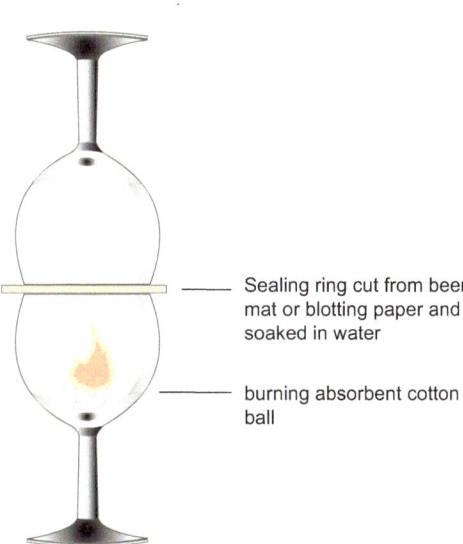

Sealing ring cut from beer mat or blotting paper and soaked in water

burning absorbent cotton ball

Fig. 8.15 Reconstruction of the "Magdeburg Hemispheres" experiment with wine glasses

What force separates the Magdeburg hemispheres?

Starting from the so-called "large Magdeburg hemispheres" of later experiments with a diameter of 60 cm and an assumed ideal vacuum and a normal air pressure of about 1000 hPa at the time of the experiment, the maximum possible force to separate the hemispheres can be calculated:

$$F = A \cdot \Delta p$$

It should be noted that only the force component perpendicular to the circular area $A = \pi R^2$, which separates the hemispheres, is effective.

For the pressure difference Δp the atmospheric pressure p_{atm} can be used under the assumption made above. This results in:

$$F = \pi R^2 p_{atm} = \pi \cdot (0.3\,\mathrm{m})^2 \cdot 10^5 \frac{\mathrm{N}}{\mathrm{m}^2} \approx 28\,\mathrm{kN}$$

What pressure difference does the "Magdeburg wine glass experiment" achieve?

From a successful experiment with small glasses, in which the lower glass "adhered" to the upper one, the measured weight force $F_G = 0.8$ N and the opening diameter of 5 cm of the wine glass can be used as a minimum force. Furthermore, a normal atmospheric pressure of 1013 hPa is again assumed here. This gives us:

$$\Delta p \geq \frac{F_G}{\pi R^2} = \frac{0.8\,\mathrm{N}}{\pi\,(0.025\,\mathrm{m})^2} = 407\,\frac{\mathrm{N}}{\mathrm{m}^2} \approx 400\,\mathrm{Pa}$$

The gas pressure achieved inside the "Magdeburg double wine glass" by cooling is therefore at most:

$$p_{inside} \leq p_{atm} - \Delta p = 1013\,\mathrm{hPa} - 400\,\mathrm{Pa} = 1009\,\mathrm{hPa}$$

Table, Cover Yourself ... Up!

Experiment 49: Tablecloth Away in 0.1 s

After many experiments, the physical wine tasting is now coming to an end. The bottles and glasses are emptied, you dear readers are educated, and it's time to clean up. Can't we also connect this with physics? Of course! Start clearing the full table with the tablecloth. It's certainly unconventional, but definitely a lively element at the end of an evening.

We have tried it and can claim to have cleared the tablecloth in less than a tenth of a second, without caring about the objects on it. Fig. 8.16 shows a sequence from a high-speed recording at 240 fps. The removal of

Fig. 8.16 The tablecloth is cleared in 0.1 s

the tablecloth took about 20 individual frames, from which this time can be estimated.

The great acceleration and the associated determination of the person pulling the tablecloth are indeed necessary for success. For this experiment, practice in a quiet room first and don't use the good glasses right away. And another important note: A tablecloth with a hemmed edge is dangerous!

How can this really short experiment be explained? For this, we have to look again at friction processes (see info box "Static and sliding friction" in Experiment 45). It is clear: Slow pulling on the tablecloth simply makes the objects on it move along. The maximum static friction force is then simply greater than the accelerating force. The crucial point is the very fast pulling movement. Upon closer inspection—e.g. a high-speed recording—it can be seen that the objects do experience a small acceleration and wobble a bit. This wobbling is especially true for objects whose center of mass is relatively far above the table surface and thus a noticeable torque is created, e.g. with a wine glass. With very fast pulling, the static friction almost instantly turns into sliding friction. The force required to accelerate is now greater than the maximum static friction force. Due to the shortness of the time for the process, the objects hardly move relative to the table, but "slide" over the tablecloth.

Let's briefly discuss Newton's 1st Law (Inertia Law) at this point, which is often invoked to explain this experiment. This law states that a body is not accelerated as long as there is no net force acting on it. But the situation on the table is obviously different. By pulling the tablecloth, the condition "net force on the body is zero" is no longer given. There are indeed additional forces acting on the objects and we can see this from the jolting and wobbling. The key point is the very short time of force application. If this time is too long, the objects move over longer distances with the tablecloth or they fall over due to the resulting torque (Fig. 8.17a). If the time span of the

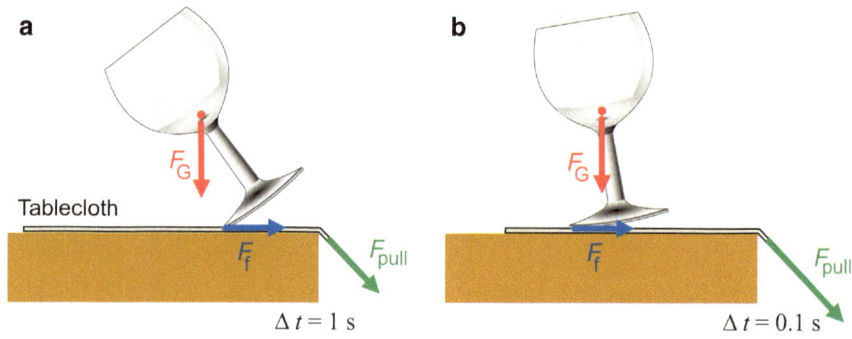

Fig. 8.17 Wine glass on pulled tablecloth; slowly pulled (**a**), pulled very quickly (**b**)

effect is very small, the objects also move, but only for this short time span. The resulting torques can't move the center of gravity out of its position above the base due to the short duration, the object wobbles, but it does not tip over (Fig. 8.17b).

Conclusion: As long as no one pulls the tablecloth, the situation can be wonderfully described with Newton's 1st Law. But it is not exciting. When the cloth is pulled, tensile and friction forces act on the glass, the sum of which is not zero! The key to clearing the table without accidents is a very short time span in which the objects on the tablecloth have no opportunity to move far or tip over.

For the minimization of the tendency of objects to move, a low friction force between the object and the tablecloth is helpful. Smooth cotton fabrics or silk support the success of this experiment. In a very simple measurement experiment, we determined the friction coefficients using the combination of a wine glass and a cotton tablecloth as an example. For a slightly more accurate measurement, the glass used in the tablecloth experiment was additionally filled with water. The ratio of tensile force to normal force does not change as a result. From the equation $F_f = \mu \cdot F_n$ the friction coefficients can be determined as the ratio of friction force to normal force.

The filled glass has a mass of 470 g, which corresponds to a gravitational force or normal force of 4.61 N. Using a spring force meter, the average values of the force required to set the glass in motion and the average values of the force required to pull the glass at a constant speed over the surface were determined from multiple repetitions. These measurements were carried out on the fixed tablecloth with the following results (Table 8.1):

For the other objects also used in the experiment (glass bottle and porcelain plate), similar values can be assumed.

Table 8.1 Experimental determination of friction coefficients ($F_n =$ 4.61 N in all cases)

Material combination	Static friction force in N	Static friction coefficient μ_{sf}	Sliding friction force in N	Sliding friction coefficient μ_{slf}
Glass/smooth cotton	0.70	0.15	0.65	0.14

A video analysis of the process shown in Fig. 8.16 provides a maximum value of 6.5 m/s^2 for the acceleration of the tablecloth. Thus, a force of approximately 1.4 N would have to act on the empty glass weighing 0.22 kg in order for it to follow the movement of the tablecloth. However, the maximum static friction force is only 0.7 N, which is why the glass cannot be accelerated to the same extent. For a filled glass, the force necessary for acceleration would even be 6 times as large as the static friction force.

Bottle Empty? Lights Out!

Experiment 50: Blowing Out Candles Through the Bottle

Now, after the clearing up has already begun, we have indeed reached the end. But wait! The bottles are still on the table and a few candles are still burning comfortably. This provides one last opportunity for physical considerations.

Before we leave the table, the burning candles should be extinguished. How about blowing them out through one of the bottles standing around? Let's try it …

This attempt is clearly successful. Of course, we cannot blow through the bottle glass, but the air still finds its way into the supposed wind shadow behind the bottle. An initial assumption seems to explain the result. The streamline image of a medium flowing perpendicular to a cylinder (our wine bottle) shows how the flow divides symmetrically at the stagnation point S_1. Behind the cylinder, the flow converges again at S_2 (Fig. 8.18). So one could be satisfied with the fact that there is no real wind shadow and therefore the candle can simply be blown out.

But the explanation of the blown out candle is not that simple! The experiment is even more astonishing—if you look very closely. If someone starts blowing behind the bottle, you can observe how the candle flame surprisingly leans towards the bottle at the beginning of the process (Fig. 8.19a),

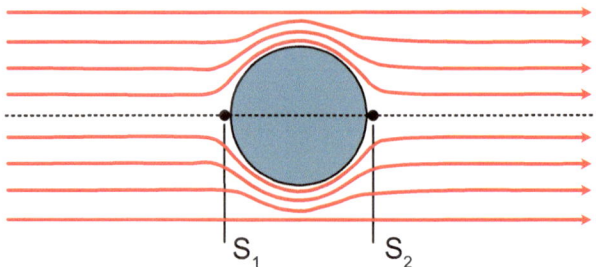

Fig. 8.18 Streamline image of an ideal flowing medium around a cylinder

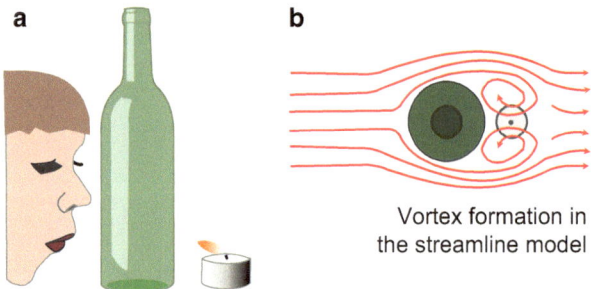

Fig. 8.19 A burning candle is seemingly blown out through the bottle. The direction of the candle flame at the beginning of the process is astonishing

then flutters around and finally goes out. Try it out, the slow-motion video function of your smartphone can show this clearly.

The unexpected inclination of the flame in the "wrong" direction raises the question of the flow conditions once again.

In the streamline image in Fig. 8.18, a so-called *ideal fluid* is assumed as the flowing medium. Strictly speaking, a body would not oppose any resistance to such a flow. Real flows like water or wind (air) differ from the ideal flows due to the friction between their layers. This changes the flow pattern behind the flowed-around body significantly. The flow then no longer follows the contour of the body and does not converge at the rear stagnation point, but separates much earlier from the surface of the obstacle. It tears off and forms vortices.

If a flow is not too strong, then in physics we speak of a low *Reynolds number*. In this case, temporally stable (stationary) vortices form behind the flowed-around cylinder, which are counter-rotating. This is exactly the effect that can be observed when not blowing out the candle flame behind a wine bottle too vigorously. Directly behind the bottle, between the vortices, there is an area with a flow direction opposite to the actual flow. These vortices cause the flame, before it goes out, to move in the backward direction towards the wine bottle (Fig. 8.19b).

Reynolds number, type of flow and vortex formation

Flows in vessels and around bodies can be described depending on various sizes. An *ideal flow* characterizes the absence of friction between the layers of a flowing medium and leads to flow situations as in Fig. 8.18.

Real flow processes are described by the *Reynolds number* (Re).

$$Re = \frac{\rho \cdot v \cdot l}{\eta}$$

Here, ρ the density of the flowing medium, v the flow velocity, l the dimension of the flowed around (or flowed through) body and η the *viscosity* (stickiness) of the flowing medium.

At very small Reynolds numbers, i.e., at very low flow velocities or high viscosity values, a flow is *laminar* and *stationary* (temporally constant). There are no vortices, the flow occurs "layer by layer" and divides symmetrically around a flowed around body.

As the Reynolds number increases, symmetric vortices form behind a flowed around cylinder (Fig. 8.19b). However, the flow remains stationary.

As the flow velocity and thus also the Reynolds number continue to increase, there is an alternating asymmetric detachment of the vortices and the formation of a so-called *Kármán vortex street*. The flow then becomes *unsteady* on the backside of the flowed around body.

With further increase in the flow velocity, the flow around the body finally begins to become *turbulent*.

Dear reader, after we have already started clearing up and now also turned off the last light, we have undoubtedly reached the end of our physical wine tasting. We, as authors, are of course very pleased about hints to the fifty experiments and further suggestions. And if you recognize the physics behind things the next time you open a bottle of wine, when clinking glasses or after the annoying spilling when pouring, but above all, if we could arouse and somewhat quench your scientific (knowledge) thirst in everyday life, if you have found pleasure in questions like those in this book, then we have achieved our goal. In this sense, we continue to wish: Cheers and stay curious!

Bibliography and Apps Used

Bibliography

Amtsblatt der Europäischen Union. VERORDNUNG (EG) Nr. 479/2008 des Rates vom 29. April 2008 über die gemeinsame Marktorganisation für Wein, zur Änderung der Verordnungen (EG) Nr. 1493/1999, (EG) Nr. 1782/2003, (EG) Nr. 1290/2005, (EG) Nr. 3/2008 und zur Aufhebung der Verordnungen (EWG) Nr. 2392/86 und (EG) Nr. 1493/1999.

Brandl, H. (2006). *Trickkiste Chemie* (2. Aufl.). Aulis-Verlag Deubner.

Denninger, G. (2013). Das Ohr trinkt mit: Schwingungen von Weingläsern und deren Klang. *Physik in unserer Zeit,44*(3), 142–146.

Gerthsen, C. (1999). *Gerthsen Physik* (20. akt. Aufl.). Springer.

Gruber, W. (2006). *Unglaublich einfach. Einfach unglaublich: Physik für jeden Tag.* Ecowin.

Kasper, L., & Vogt, P. (2020). Corkscrewing and speed of sound: A surprisingly simple experiment. *The Physics Teacher,58,* 278–279.

Kasper, L., & Vogt, P. (2022). Physik im Weinkeller. Durchs Glas geschaut: Rotwein als Farbfilter. *Physik in unserer Zeit, 1*(53), 43.

Levine, H., & Schwinger, J. (1948). On the radiation of sound from an unflanged circular pipe. *Physical review, 73*(4), 383.

Liger-Belair, G., Cordier, D., et al. (2017). Unveiling CO_2 heterogeneous freezing plumes during champagne cork popping. *Scientific Reports 7, 10938,* 1–12.

Lüders, K., & von Oppen, G. (2008). *Bergmann/Schaefer. Lehrbuch der Experimentalphysik, Band 1, Mechanik, Akustik, Wärme.* De Gruyter.

Mamola, K. C., & Pollock, J. (1993). The breaking broomstick demonstration. *The Physics Teacher, 31*(4), 230–233.

L. Kasper and P. Vogt, *Uncorking the Physics of Wine*,
https://doi.org/10.1007/978-3-662-68759-8

Mangold, K., Shaw, J. A., & Vollmer, M. (2015). Rotwein zu Wasser. Infrarotfotografie mit kommerziellen Digitalkameras. *Physik in unserer Zeit, 1*(46), 12–16.

Monteiro, M., Marti, A., Vogt, P., Kasper, L., & Quarthal, D. (2015). Measuring the acoustic response of Helmholtz resonators. *The Physics Teacher,53,* 247–249.

Myhrvold, N., Young, C., & Bilet, M. (2011). *Modernist cuisine: The art and science of cooking*. The Cooking Lab.

Nickolaus, P. (2018). *Einfluss von Sauerstoff auf die Polymerisation von Rotweinpigmenten*. Universität Kaiserslautern.

Reclari, M., Dreyer, M., Tissot, S., Obreschkow, D., Wurm, F. M., & Farhat, M. (2014). Surface wave dynamics in orbital shaken cylindrical containers. *Physics of Fluids,26*(5), 052104.

Schlichting, H. J., & Ucke, C. (1995). Es tönen die Gläser. *Physik in unserer Zeit, 26*(3), 138–139.

Schmidt, W. (1899). *Herons von Alexandria Druckwerke und Automatentheater*. Teubner.

Sigloch, H. (2003). *Technische Fluidmechanik*. Springer.

Sun, T. (2018). *Numerical investigation of mass transfer at non-miscible interfaces including Marangoni force*. Technische Universität München.

Tornaría, F., Monteiro, M., & Marti, A. C. (2014). Understanding coffee spills using a smartphone. *The Physics Teacher,52,* 502–503.

Trendelenburg, F. (1950). Einführung in die Akustik (2. Aufl.). Springer.

Vogt, P., & Kasper, L. (2015). Abschätzung des Drucks in Sektflaschen mithilfe einer Hochgeschwindigkeitsvideoanalyse. *Naturwissenschaften im Unterricht Physik,146,* 49–50.

Vogt, P., & Kasper, L. (2021a). Das fallende Weinglas – Ein überraschender Freihandversuch zum Thema „Rotation". *Naturwissenschaften im Unterricht Physik,181,* 49–50.

Vogt, P., & Kasper, L. (2021b). Die akustische Schwebung mit Weingläsern: Quantitative Analyse mit dem Smartphone. *Naturwissenschaften im Unterricht Physik, 183/184,* 97–98.

Vogt, P., & Kasper, L. (2022). Physik im Weinkeller. Geschüttelt oder gerührt? – Geschleudert! *Physik in unserer Zeit, 2*(53), 100.

Vogt, P., Kasper, L., & Müller, A. (2014). Physics^{2}Go! Neue Experimente und Fragestellungen rund um das Messwerterfassungssystem Smartphone. *PhyDid B – Didaktik der Physik – Beiträge zur DPG-Frühjahrstagung*.

Vogt, P., Kasper, L., & Burde, J.-P. (2015). The sound of church bells: Tracking down the secret of a traditional arts and crafts trade. *The Physics Teacher,53,* 438–439.

Vogt, P., Kasper, L., & Burde, J.-P. (2016). More sound of church bells: Authors' correction. *The Physics Teacher,54,* 52–53.

Apps

RWTH Aachen (Aachen University). (2016). phyphox. available for iOS: https://ogy.de/phyphox-iOS – for Android: https://ogy.de/phyphox-Android.

Sinusoid Pty Ltd. (2020). Audio Kit. available for iOS: https://ogy.de/Audio-Kit.

Wagner, A. (2011). Spektroskop. available for iOS: https://ogy.de/Spektroskop.

Ziegler, M. (2014). Spaichinger Schallanalysator. available for iOS: https://ogy.de/Schallanalysator-iOS – for Android: https://ogy.de/Schallanalysator-Android.

Index

Aberration, spherical 72
Absorption filters 77
Acceleration 17–19, 124–127, 157, 159
Acetaldehyde 26, 39
Acetic acid 38
Adhesive force 86, 122
Aerating 25, 34, 36
Aeration vi, 28, 30, 32, 33, 35, 36
Agraffe 16
Air pressure 10, 22, 84, 86, 87, 105, 109, 153, 155, 156
Air pump 153
Alcohol content 11, 39, 90, 97, 98, 100, 101
Angular momentum 130
Anomaly of water 12, 96
Anthocyanins 39, 76, 77
Antinode of motion 2
Archimedes number 140
Archimedes' principle 150, 151
Aroma v, 26, 28, 34, 36, 38, 92
Astigmatism 73
Astringency 26

Balance 116
Bang 44
Bar (unit) 9
Beat
    acoustic 46–48
    complete 47
    incomplete 47
Beat frequency 46–48
Bernoulli equation 32
Bicycle pump 8, 9
Blanc de noirs 69, 78–81
Blood alcohol concentration 140, 142, 143
Blue-Bottle-Experiment 26
Bocksbeutel 58
Boiling 10, 16
Boiling point 94, 95
Boiling temperature 98
Bottle holder 119, 120
Bottle neck resonator 2
Boyle, Robert 115
Boyle-Mariotte law 114, 115
Browning 26
Buoyancy force 139
Buoyant force 96, 149–151

Cambrer 91
Carbon dioxide 16, 19
Cavity resonator 57
Center of gravity 103, 116, 118–120, 122, 158
Center of mass 118
Centripetal force 28, 125
Church bell 44
Clay 93, 94
Cobbler's ball 69–71
Coffee Mug Model 61
Cohesion force 152
Cohesive force 86, 94, 100, 122, 123
Color filter 76
Color mixing, subtractive 76
Conservation of angular momentum 130
Conservation of energy 32
Continuity equation 31
Convection 90, 139
    chemical 139
    forced 140
    free 139
Cork oak 148
Cork pulling 4, 15
Corkscrew 6, 10, 13, 22, 65, 116, 117, 121, 128, 149, 150
Corkscrewing vi, 23, 24
Corkscrew rule 121
Cup of justice 107

Decanting vi, 25, 29, 33, 34, 36
Decay time 44
Density 31, 32, 87, 96, 97, 106, 112, 139, 148, 150, 151, 161
Digestif 98
Distillation 98
Diving bell 113–115

End correction 3
Equilibrium 118–120, 123
    stable 117
    static 118
Ethanol 11, 12, 38, 39, 89, 90, 94, 97, 98, 152
Evaporation 12, 13, 90, 92, 94, 95
Excess pressure 9, 16, 18, 19
Excitation frequency 52
Extinction 81

Fast Fourier Transform 55
Fermentation 16, 38, 39, 78, 79
Flow
    ideal 161
    laminar 161
    stationary 161
    turbulent 161
Flow velocity 31, 32, 139, 161
Focal length 72
Focal line 74, 75
Focal point 70, 72, 74
Force 7
Force impulse 133
Force meter 18
Force of gravity 138
Force-transforming devices 22
Fourier analysis 44
Fourier theorem 44
Frapper 91
Freezing point 98
Frequency spectrum 44, 45, 49, 53, 55
Friction 2, 22, 53, 60, 144–146, 148, 149, 160
    Kinetic friction 146
    Sliding friction 63, 145, 146, 157
    Static friction 145, 146, 157
Frictional force 8, 9, 18, 146

Friction coefficient 63, 146–148, 158, 159
Friction force 20, 129, 147, 158
Fundamental frequency 43–45, 57–59, 63, 65, 66, 141, 142
Fundamental tone 42, 43, 49
Fundamental vibration 2, 61, 62

General gas equation 114
Glass harmonica 63
Golden Rule of Mechanics 22
Gravitational force 28, 100, 109, 127, 151, 158
Gravitational pressure 109, 110, 114, 150
Gravity 32, 89, 90, 120, 121, 151

Heat of evaporation 92
Heat of vaporization 94
Helmholtz resonator 56, 58, 59, 64
Heron of Alexandria v, 103, 104
Hydrophilicity 84
Hydrostatic paradox 112
Hydrostatic pressure 105, 106
Hyperdecanting 33

Ideal gas equation 14
Impulse 7, 133
Inertia law 157
Infrared blocking filters 79
Infrared pass filter 80
Infrared photography 79, 80
Interfacial tension 90, 100
Interference filters 77

Kármán vortex street 161
Kinetic friction force 63

Law of conservation of angular momentum 129
Law of conservation of energy 128
Law of continuity 31, 32
Law of energy conservation 109
Lens
    Cylindrical lens 73–75
    Spherical lens 69–73, 75
Lens error 72
Leukomethylene blue 27
Lever law 22
Lever
    one-sided 22
    two-sided 22
Light, infrared 79
Liquid expansion 12
Longitudinal wave 135
Lotus effect 85

Mach, Ernst 104
Magdeburg hemispheres 153, 155, 156
Manometer 9
Marangoni, Carlo 89
Marangoni convection 90
Marangoni effect 90
Marangoni number 90
Mariotte, Edme 115
Melting 95
Meniscus 152
Methylene blue 26, 27
Momentum 7, 130
Mouth correction 4
Mulled wine 13

N

Natural frequency 49–52, 63, 143
Negative pressure 38
Newton, Isaac 19
Newton (unit) 19
Nodal line 61
Node of motion 2
Noise, white 53, 55, 56
Noise 44
Note 65

O

Orbital speed 129, 130
Organ pipe 2
Oscillator 143
Oscillogram 44–46
Overtone 2, 42–44
Overtones 61
Oxidation 26, 27, 30, 38, 39

P

Palatinate tube 53–55, 73, 124
Pascal (unit) 9
Period 143
Pipe, stopped 2
Pirouette effect 130
Pitch 62, 63, 141
Pivot point 120
Polyphenols 26
Pop sound 1, 2, 16
Pressure 1, 8, 9, 12, 14–21, 31, 32, 37,
    38, 60, 86, 87, 105, 106, 109,
    111–115, 148, 150
  atmospheric 15, 20, 87, 114
  hydrostatic 87
  stagnation pressure 32
  static 32
Pressure antinode 2
Pressure node 2, 3

Principle of communicating tubes 108,
    112
Propagation speed 2, 133, 135, 141,
    144
Pulling the cork 146
Pythagoras of Samos 107
Pythagorean cup 107

Q

Quality sparkling wine 17

R

Range of tones 65
Rayleigh scattering 79, 81
Refraction, optical 71
Refractive index 71–73
Resonance vi, 49, 52, 64
Resonance catastrophe 49, 52
Resonance curve 50, 51
Resonance frequency 2, 3, 5, 53, 56,
    57, 61, 63
Resonance tube 2, 3
Resonator 57, 64, 65
Resting position 117
Rest position 52, 129
Reynolds number 160, 161
Rise height, capillary 100, 101

S

Screw rule 22, 121
Siphon 108, 109
Sliding friction force 7, 146, 159
Sound 2, 42–46, 55, 60, 61, 63
Sound pressure level 49
Sound propagation 22, 67
Sound speed 6, 133
Sound waves 46, 49, 143, 144
Spectral line 44

Speed 71, 146
Speed of sound 1–5, 43, 53–58, 66, 67, 135, 141, 142
Spill 124
Spilling 125
Spindle corkscrew 23
Spritzer 73, 110
State change 15, 20, 21
  adiabatic 21
  isochoric 21
  isothermal 21, 115
Static friction force 11, 17, 18, 63, 146, 147, 157, 159
Stick-slip effect 60, 63, 64
Streamline image 159, 160
Strutt, John William 81
Sublimation 20
Sulfur dioxide 39
Surface, free 127
Surface tension 85, 88–90, 100, 101, 152
Swirl 89
Swirling 26–29, 89
System center of gravity 116, 118, 120

Tannin 26, 34
Tantalus cup 107, 108, 110
Teapot effect 84
Temperature, absolute 67
Thermal conductivity 118, 148
Thermal expansion 11, 14
Thermal expansion coefficient 11–13
Tidal friction 130
Timbre 44
Tone 44, 46, 48, 49, 58, 63, 65, 66
Tone generator 49, 51
Torque 118–121, 130, 157, 158
Transmittance 77

Transverse wave 135
Tubes, communicating 111

Uncorking 1–6, 8–10, 13, 14, 18, 53, 144

Vacuum 37, 109, 156
Vapor pressure 87
Velocity 7
Venturi nozzle 31, 33
Vibration belly 61, 135
Vibration
  forced 52
  free 52
Vibration node 61, 135
Vinometer 100
Viscosity 90, 161
von Guericke, Otto 37, 153, 154
von Helmholtz, Hermann 55, 57
Vortices 160, 161

Wave crest 61
Wave peak 29
Wave trough 61
Weight force 87, 96, 106, 112, 120, 122, 128, 146, 150, 155, 156
Wine automaton 104, 105
Wine cooler 92–94
Wine-water mixing machine 110